APPLIED LINEAR REGRESSION is a guide to the methodology for fitting linear models to data, and includes many recent advances in the field. The central themes are model building, assessing fit and reliability, and drawing conclusions. Standard regression topics, such as least squares estimation and confidence and testing procedures, are carefully presented and illustrated through the use of examples. Residual analysis, choice of transformations for variables, and finding influential cases are covered in great detail. Model building, including the definition of independent and dummy variables, and subset selection are discussed. The reasons for selecting a subset are emphasized; for the computations, analysis of all possible regressions is advocated rather than stepwise algorithms, and both are illustrated by example. Additional topics include statistics for investigating the sensitivity of results to measurement or rounding error in the independent variables, special considerations for prediction models, a discussion of modifications necessary for the analysis of incomplete data, and non-least squares estimation methods.

While not tied to any specific computer program, this book is intended for the reader who expects to use a computer to fit linear regression models to data. Computational considerations are discussed in sufficient detail to suggest numerically stable methods for those interested in writing computer software. Sources of software are included.

This book is a primary statistical tool for statisticians, computer scientists, econometricians, applied mathematicians, and social and biological scientists. Economists and forecasters will find this a valuable reference to model building and validation, collinearity, and prediction. Professors and students will find the first eight chapters a detailed treatment of the subject. Theoretical problems are provided for students with the necessary statistical background, while problems that require analysis of data are suitable for a wider audience. Most of the problems and almost all of

(cont'd on back flap)

D0464202

sing real miliarity ncepts: testing, estimation, and distributions for random variables. The required background in matrix algebra is covered in the appendix.

Inference

, OAKES, VANDAELE, and for Comparative Studies ic Processes with Applications

nd Systems for Health
Statistical Data
lodels for Social Processes,

Statistical Techniques for Man-

Estimation in Engineering and

Regression Diagnostics: Iden-
s of Collinearity
istical Analysis in Chemistry

stic Processes
f Time Series: An Introduction
cientist
ry Operation: A Statistical

atistics for Experimenters: An
ysis, and Model Building
ics: A Biomedical Introduction
d Methodology in Science and

d Science
iods for Data Analysis
ssion Analysis by Example
tary Decision Theory
ynamic Economic Systems
I • Basic Statistics with Busi-

Third Edition
al Designs, *Second Edition*
ic Statistics, *Second Edition*

on for Analysis in the Health

to Industrial Experimentation
uations to Data: Computer
ıd Edition

ss Research
continued on back

APPLIED LINEAR REGRESSION

Applied Linear Regression

SANFORD WEISBERG

University of Minnesota
St. Paul, Minnesota

JOHN WILEY & SONS
New York · Chichester · Brisbane · Toronto

Library of Congress Cataloging in Publication Data:

Weisberg, Sanford, 1947–
  Applied linear regression.

  (Wiley series in probability and mathematical
statistics)
  Bibliography: p.
  Includes index.
  1. Regression analysis. I. Title.

QA278.2.W44   519.5'36   80-10378
ISBN 0-471-04419-9

Printed in the United States of America

10 9 8 7 6 5 4 3 2

To my parents and to Carol

PREFACE

Linear regression analysis consists of a collection of techniques used to explore relationships between variables. It is interesting both theoretically because of the elegance of the underlying theory, and from an applied point of view, because of the wide variety of uses of regression that have appeared, and continue to appear every day. In this book, regression methods, used to fit models for a dependent variable as a function of one or more independent variables, are discussed for the reader who wants to learn to apply them to data. The central themes are building models, assessing fit and reliability, and drawing conclusions. If used as a textbook, it is intended as a second or third course in statistics. The only definite prerequisites are familiarity with the ideas of significance tests, p-values, confidence intervals, random variables, estimation of parameters, and also with the normal distribution, and distributions derived from it, such as Student's t, and the F, and χ^2. Of course, additional knowledge of statistical methods or linear algebra will be of value.

The book is divided into 11 chapters. Chapters 1 and 2 provide fairly standard results for least squares estimation in simple and multiple regression, respectively. The third chapter is called "Drawing Conclusions" and is about interpreting the results of the methods from the first two chapters. Also, a discussion of the effects of independent variables that are imperfectly measured is given. Chapter 4 presents additional results on least squares estimation. Chapters 5 and 6 cover methods for studying the lack of fit of a model, checking for failures of assumptions, and assessing the reliability of a fitted model. In Chapter 5, theoretical results for the necessary statistics are given, since these will be unfamiliar to many readers, while Chapter 6 covers graphical and other procedures based on these statistics, as well as possible remedies for the problems they uncover. In Chapter 7, the topics covered are relevant to problems of model building, including dummy variables, polynomial regression, and principal components. Then, Chapter 8 provides methods for selecting a model based on a subset of variables. In Chapter 9, special considerations when regression methods are to be used to make predictions are discussed. In each of these chapters, the methods discussed are illustrated by examples using real data.

The last two chapters are shorter than the earlier ones. Chapter 10 gives guidelines for analysis of partially observed or incomplete data. Finally, in Chapter 11, alternatives to least squares estimates are discussed.

Several of the chapters have associated appendixes that have been collected at the end of the text, but are numbered to correspond to the chapters. For example: Appendix 1A.2 is the second appendix for Chapter 1. The chapters are ordered for a semester or quarter course on linear regression, and Chapters 1 to 8 make up a rigorous one-quarter course.

Homework problems are provided for each of the first nine chapters. The theoretical problems are intended only for students with the necessary statistical background. Problems that require analysis of data are intended for everyone. Some of these have been left vague in their requirements, so that they can be varied according to the interests of the students. Most of the problems use real data and can be approached in many ways.

Computers. The growth of the use of regression methods can be traced directly to wider availability of computers. While this book is not intended as a manual for any specific computer program, it is oriented for the reader who expects to use computers to apply the techniques learned. High quality software for regression calculations is available, and references to the necessary sources are in the text, in the homework problems, and in the appendixes.

Acknowledgments. I am grateful to the many people who have commented on early drafts of the book, supplied examples, or through discussion have clarified my own thoughts on the topics covered. Included in this group are Christopher Bingham, Morton Brown, Cathy Campbell, Dennis Cook, Stephen Fienberg, James Frane, Seymour Geisser, John Hartigan, David Hinkley, Alan Izenman, Soren Johansen, Kenneth Koehler, David Lane, Kinley Larntz, John Rice, Donald Rubin, Wei-Chung Shih, G. W. Stewart, Douglas Tiffany, Carol Weisberg, Howard Weisberg, and an anonymous reader. Also, I wish to thank the production staff at the University of Minnesota, Naomi Miner, Sue Hangge, Therese Therrien, and especially Marianne O'Brien, whose expert assistance made completion of this work a reality.

During the writing of this book, I have benefited from partial support from a grant from the U.S. National Institute of General Medical Sciences. Additional support for computations has been provided by the University Computer Center, University of Minnesota.

SANFORD WEISBERG

St. Paul, Minnesota
February 1980

CONTENTS

APPLIED LINEAR REGRESSION

1

SIMPLE LINEAR REGRESSION

Regression is used to study relationships between measurable variables. Linear regression is used for a special class of relationships, namely, those that can be described by straight lines, or by generalizations of straight lines to many dimensions. These techniques are applied in almost every field of study, including social sciences, physical and biological sciences, business and technology, and the humanities. As illustrated by the examples in this book, the reasons for fitting linear regression models are as varied as are the applications, but the most common reasons are description of a relationship and prediction of future values.

Generally, regression analysis consists of many steps. To study a relationship between a number of variables, data are collected on each of a number of units or cases on these variables. In the regression models studied here, one variable takes on the special meaning of a response variable, while all of the others are viewed as predictors of the response. It is often convenient, and sometimes accurate, to view the predictor variables as having values set by the data collector, while the response is a function of those variables. A hypothesized model specifies, except for a number of unknown parameters, the behavior of the response for given values of the predictors. The model generally will also specify some of the characteristics of the failure to provide exact fit through hypothesized error terms. Then, the data are used to obtain estimates of unknown parameters. The method of estimation studied in this book is *least squares*, although there are in fact many estimation procedures. The analysis to this point is called *aggregate analysis*, since the main purpose is to combine the data into aggregates and summarize the fit of a model to the data. The next, and equally important, phase of a regression analysis is called *case analysis*, in which the data are used to examine the suitability and usefulness of the fitted model for the relationship studied. The results of case analysis will often lead to modification of the original prescription for a fitted model, and cycling back to the aggregate analysis after modifying the data or assumptions is often necessary.

The topic of this chapter is simple regression, in which there is a single response and a single predictor. Of interest will be the specification of an appropriate model, discussion of assumptions, the least squares estimates, and testing and confidence interval procedures.

Example 1.1 Forbes' data

In the 1840s and 1850s a Scottish physicist, James D. Forbes, wanted to be able to estimate altitude above sea level from measurement of the boiling point of water. He knew that altitude could be determined from atmospheric pressure, measured with a barometer, with lower pressures corresponding to higher altitudes. In the experiments described, he studied the relationship between pressure and boiling point. His interest in this problem was motivated by the difficulty in transporting the fragile barometers of the 1840s. Measuring the boiling point would give travelers a quick way of estimating altitudes.

Forbes collected data in the Alps and in Scotland. After choosing a location, he assembled his apparatus, and measured pressure and boiling point. Pressure measurements were recorded in inches of mercury, adjusted for the difference between the ambient air temperature when he took the measurements and a standard temperature. Boiling point was measured in degrees Fahrenheit. The data for $n = 17$ locales are reproduced from an 1857 paper in Table 1.1 (Forbes, 1857).
On reviewing the data, there are several questions of potential interest. How are pressure and boiling point related? Is the relationship strong or weak? Can we predict pressure from temperature, and if so, how well?

Forbes' theory suggested that over the range of observed values the graph of boiling point versus the *logarithm* of pressure yields a straight line. Following Forbes, we take logs to the base 10, although the base of the logarithms is irrelevant for the statistical analysis. Since the logs of the pressures do not vary much, with the smallest

Table 1.1   Forbes' data, giving boiling point (°F) and barometric pressure (inches of mercury) for 17 locations in the Alps and in Scotland.

Case Number	Boiling Point (°F)	Pressure (in. Hg)	Log(Pressure)	100 × Log(Pressure)
1	194.5	20.79	1.3179	131.79
2	194.3	20.79	1.3179	131.79
3	197.9	22.40	1.3502	135.02
4	198.4	22.67	1.3555	135.55
5	199.4	23.15	1.3646	136.46
6	199.9	23.35	1.3683	136.83
7	200.9	23.89	1.3782	137.82
8	201.1	23.99	1.3800	138.00
9	201.4	24.02	1.3806	138.06
10	201.3	24.01	1.3805	138.05
11	203.6	25.14	1.4004	140.04
12	204.6	26.57	1.4244	142.44
13	209.5	28.49	1.4547	145.47
14	208.6	27.76	1.4434	144.34
15	210.7	29.04	1.4630	146.30
16	211.9	29.88	1.4754	147.54
17	212.2	30.06	1.4780	147.80

being 1.318 and the largest being 1.478, we shall multiply all the values of log(pressure) by 100, as given in column 5 of Table 1.1. This will avoid studying very small numbers, without changing the major features of the analysis.

A useful way to begin a regression analysis is by drawing a graph of one variable versus the other. This graph, called a *scatter plot*, can serve both to suggest a relationship, and to demonstrate possible inadequacies of it. Scatter plots can be drawn on ordinary graph paper. The x axis (or horizontal axis) is usually reserved for the variable that is to be the predictor or describer, or independent variable. In Forbes' data this is the boiling point. The y axis or the vertical axis is usually for the quantity to be modeled or predicted, often called the response or the dependent variable. In the example, the values for the y axis are 100 × log(pressure). For each of the n pairs (x, y) of values in the data, a point is plotted on the graph. Although easily produced with pencil and paper, most computer programs for regression analysis will produce this plot.

The overall impression of the scatter plot for Forbes' data (Figure 1.1) is that the points generally, but not exactly, fall on a straight line (the line drawn in Figure 1.1 will be discussed later). This suggests that the relationship between the two variables may be described (at least as a first approximation) by specifying an equation for a straight line.

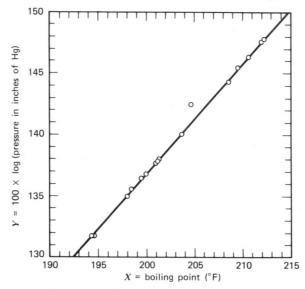

Figure 1.1   Scatter plot for Forbes' data.

As we progress through this chapter, the methods studied will be applied to these data.

1.1   Building a simple regression model

In simple regression, the relationship between two quantities, say X and Y, is studied. First, we hope that the relationship can be described by a straight line. For this to be reasonable, we may need to transform the scales of the quantities X and/or Y, as was done in Forbes' data, where pressure was transformed to log(pressure). In this chapter, observed values of the quantities X and Y are denoted by subscripted lower case letters: (x_i, y_i) are the observations on X and Y for the ith case in the study. The major features of the simple regression model are given here. A more formal approach is given in Appendix 1A.1.

Equation of a straight line. A straight line relating two quantities Y and X can be described by the equation

$$Y = \beta_0 + \beta_1 X \tag{1.1}$$

where β_0 is called the *intercept*, and corresponds to the value of Y when $X = 0$ (and is therefore the point where the line intercepts the y axis), and β_1 is called the *slope*, giving the change in Y per unit change in X (see Figure 1.2). The numbers β_0 and β_1 are called *parameters*, and, as they range over all possible values, they give all possible straight lines. In statistical applications of straight line modeling, these parameters are generally unknown, and must be estimated using the data. The difference between estimates of parameters computed from data and the actual, though unknown, values of the parameters is very important, since the data provide information about the parameters, not their actual values.

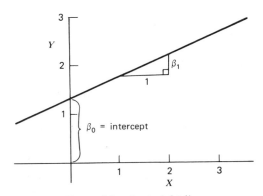

Figure 1.2   A straight line.

Errors. Real data will almost never fall exactly on a straight line. The differences between the values of the response obtained and the values given by the model (e.g., for simple regression, the observed values of Y minus $(\beta_0 + \beta_1 X)$) are called statistical *errors*. This term should not be confused with its synonym in common usage, "mistake." Statistical errors are devices that account for the failure of a model to provide exact fit. They can have both fixed and random components. A fixed component of a statistical error will arise if the proposed model, here a straight line, is not exactly correct. For example, suppose the true relationship between Y and X is given by the solid curve in Figure 1.3, and suppose that we incorrectly propose a straight line, shown as a dashed line, for this relationship. By modeling the relationship with a straight line rather than the appropriate curve, a fixed error, sometimes called the lack of fit error, is the vertical distance between the straight line and the correct curve. For the standard linear regression theory of this chapter, we assume that the lack of fit components to the errors are negligible.

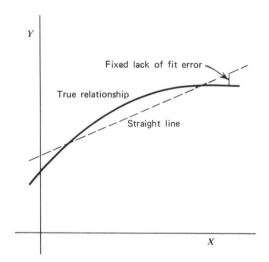

Figure 1.3   Approximating a curve by a straight line.

For the purposes of this chapter, the random component to the errors is more important. The random component can have several sources. Measurement errors (for now, consider only errors in Y, not X) are almost always present, since few quantitative variables can be measured with perfect accuracy. The effects of variables not explicitly included in the model can contribute to the errors. For example, in Forbes' experiments wind speed may have small effects on the atmospheric pressure, contributing to the variability in the observed values. Also, random errors due to natural variability occur.

Let e_i be the value of the statistical error for the ith case, $i = 1, 2, \ldots, n$. Assuming that the fixed component of the errors is negligible, the e_i have zero mean, $E(e_i) = 0$, $i = 1, 2, \ldots, n$. (See Appendix 1A.2 if the symbols $E(\ )$, var$(\ )$, and corr$(\ , \ )$ are unfamiliar.) An additional convenient assumption is that the errors are mutually uncorrelated (written in terms of the covariance operator, as cov$(e_i, e_j) = 0$, for all $i \neq j$), and have common, though generally unknown, variance var$(e_i) = \sigma^2$, $i = 1, 2, \ldots, n$. Heuristically, uncorrelated means that the value of one of the errors does not depend on or help determine the value of any other error. Little generality is lost if the word *independent* is substituted for *uncorrelated*. An even stronger assumption that is occasionally needed is that the errors are normally distributed. This assumption is usually needed only to obtain confidence intervals and in testing procedures. In this book, we will generally not assume normality except where it is explicitly required. All of the above assumptions can be written in the compact form

$$e_i \sim \text{NID}(0, \sigma^2) \qquad i = 1, 2, \ldots, n$$

which is read as the e_i are normally and independently distributed with zero mean and common variance σ^2. In any practical problem, the assumptions made in this section must be examined. This topic is discussed in later chapters.

The simple regression model. We have already defined X and Y to be the independent variable and the dependent variable, respectively, with observed values (x_i, y_i) of X and Y for $i = 1, 2, \ldots, n$. For Forbes' data, the x_i's are given in the second column of Table 1.1 and the y_i's are given in the fifth column. For example, $x_3 = 197.9$ and $y_{12} = 142.44$. We again define e_i to be the statistical error for the ith case, $i = 1, 2, \ldots, n$. The simple linear regression model specifies the following:

$$y_i = \beta_0 + \beta_1 x_i + e_i \qquad i = 1, 2, \ldots, n \tag{1.2}$$

$$\text{with } E(e_i) = 0$$

$$\text{var}(e_i) = \sigma^2$$

$$\text{cov}(e_i, e_j) = 0, i \neq j$$

In words, the model says that the observed value y_i can be determined from the value of x_i through the specified equation, except that e_i, an unknown random quantity, is added on. The three quantities β_0, β_1, and σ^2 are unknown. The e_i's are unobservable quantities introduced into the model to account for the failure of the observed values to fall exactly on a single straight line. Only the x_i's and the y_i's are observed and these data are used to obtain estimates of the unknown parameters, namely, β_0, β_1, and σ^2.

1.2   Least squares estimation

Many methods have been suggested for obtaining estimates of parameters in a model. The one discussed here is called *least squares*, in which parameter estimates are chosen to minimize a quantity called the residual sum of squares.

Notation. The distinction between parameters and estimates of parameters (statistics) is critical to the use and understanding of statistical models. To keep this distinction clear, parameters are denoted by lower case Greek letters, usually α, β, γ, and σ, and estimates of parameters denoted by putting a "hat" over the corresponding Greek letter; thus, for example, $\hat{\beta}_1$ (read "beta one hat") is the estimator of β_1. Similarly, $\hat{\sigma}^2$ is the estimator of σ^2. Although the e_i's are not parameters in the usual sense, we shall use the same hat notation to specify the observed fitting errors or *residuals*: the residual for the ith case, denoted by $\hat{e}_i$, is given by the equation

$$\hat{e}_i = y_i - (\hat{\beta}_0 + \hat{\beta}_1 x_i) \qquad i = 1, 2, \ldots, n \tag{1.3}$$

which should be compared to the equation for the statistical errors,

$$e_i = y_i - (\beta_0 + \beta_1 x_i) \qquad i = 1, 2, \ldots, n \tag{1.4}$$

The difference between the e_i's and the $\hat{e}_i$'s is important, as the residuals are observable (and will be used to check assumptions), while the statistical errors are not observable.

An additional extension of the hat notation is used to identify *fitted values* determined by the estimated regression equation. Thus the ith fitted value $\hat{y}_i$ is given by the equation

$$\hat{y}_i = \hat{\beta}_0 + \hat{\beta}_1 x_i \qquad i = 1, 2, \ldots, n \tag{1.5}$$

By comparing (1.5) to (1.3), note that $\hat{e}_i = y_i - \hat{y}_i$.

All least squares computations can be done using only a few summary statistics computed from the data, namely, the sample averages, corrected sums of squares, and corrected cross products. For reference, the definitions of all of these quantities are given in Table 1.2. They have all been defined by subtracting the sample average from each of the values before squaring or taking cross products. Appropriate alternative formulas for computing the corrected sums of squares and cross products from uncorrected sums of squares and cross products are also given in the table. Using the uncorrected sums of squares is convenient for computing on a calculator, since many have facilities for accumulating both $\sum x_i$ and $\sum x_i^2$ in a single pass through the data. A formula such as $SXX = \sum x_i^2 - (\sum x_i)^2/n$ can be used to obtain the corrected sum of squares from the uncorrected. However, if computations are to be done on a computer, using the uncorrected sums of squares may lead to severe round-off errors, as illustrated in Appendix 1A.5.

Table 1.2 also lists definitions for the "usual" univariate and bivariate summary statistics, namely, the sample averages ($\bar{x}$, $\bar{y}$), sample variances (s_X^2, s_Y^2), and estimated covariance and correlation (s_{XY}, r_{XY}). Note that the

Table 1.2   Definitions of symbols[a]

Quantity	Definition and Alternative Forms	Description
$\bar{x}$	$\sum x_i / n$	Sample average for the x_i's
$\bar{y}$	$\sum y_i / n$	Sample average for the y_i's
SXX	$\sum (x_i - \bar{x})^2 = \sum x_i^2 - (\sum x_i)^2 / n$ $= \sum x_i^2 - n(\bar{x})^2$	Corrected sum of squares for the x_i's
s_X^2	$SXX/(n-1)$	Sample variance of the x_i's
s_X	$\sqrt{SXX/(n-1)}$	Sample standard deviation
SYY	$\sum (y_i - \bar{y})^2 = \sum y_i^2 - (\sum y_i)^2 / n$ $= \sum y_i^2 - n(\bar{y})^2$	Corrected sum of squares for the y_i's; also called the total sum of squares
s_Y^2	$SYY/(n-1)$	Sample variance of the y_i's
s_Y	$\sqrt{SYY/(n-1)}$	Sample standard deviation
SXY	$\sum (x_i - \bar{x})(y_i - \bar{y})$ $= \sum x_i y_i - (\sum x_i)(\sum y_i)/n$ $= \sum x_i y_i - n\bar{x}\bar{y}$	Corrected sum of cross products
s_{XY}	$SXY/(n-1)$	Sample covariance
r_{XY}	$SXY/\sqrt{(SXX)(SYY)} = s_{XY}/s_X s_Y$	Sample correlation

[a] The symbol $\sum$ is shorthand for $\sum_{i=1}^{n}$, which means "add for all values of i between 1 and n."

"hat" rule described above would suggest that different symbols should be used for these quantities; for example, $\hat{\rho}_{XY}$ might be more appropriate for the sample correlation if the population correlation is ρ_{XY}. However, this inconsistency is deliberate, because in many regression situations, these statistics are not estimates of population parameters. For example, in Forbes' experiments, data were collected at 17 selected locations. As a result, the sample variance of boiling points, which can be shown to be $s_X^2 = 33.17$, cannot be expected to be an estimate of any meaningful population variance. Similarly, r_{XY} depends as much on the method of sampling as it does on the population value ρ_{XY}, should such a population value make sense.

However, these usual sample statistics are often presented and used in place of the corrected sums of squares and cross products, so alternative formulas are given using both sets of quantities.

Least squares criterion. Once $\hat{\beta}_0$ and $\hat{\beta}_1$ have been obtained, the fitted values are $\hat{y}_i = \hat{\beta}_0 + \hat{\beta}_1 x_i$. The residuals $\hat{e}_i = y_i - \hat{y}_i$ represent the observed fitting errors. In least squares, we choose $\hat{\beta}_0$ and $\hat{\beta}_1$ to make the residual sum of squares, *RSS*, as small as possible, where

$$RSS = \sum \hat{e}_i^2 = \sum (y_i - \hat{y}_i)^2 = \sum \left[y_i - (\hat{\beta}_0 + \hat{\beta}_1 x_i) \right]^2 \tag{1.6}$$

Note that least squares is a purely mathematical formulation that does not depend on any assumptions concerning the e_i's. Least squares estimates can be computed even if the regression model is inappropriate for the data studied.

The least squares estimates can be found in many ways, one of which is outlined in Appendix 1A.3. They are given by the expressions

$$\hat{\beta}_1 = \frac{SXY}{SXX} = r_{XY} \frac{s_Y}{s_X} = r_{XY} \left(\frac{SYY}{SXX} \right)^{1/2}$$

$$\hat{\beta}_0 = \bar{y} - \hat{\beta}_1 \bar{x} \tag{1.7}$$

The several forms for $\hat{\beta}_1$ are all equivalent.

Sometimes it is convenient to write the simple linear regression model in a different form that is a little easier to manipulate. Taking equation (1.2), and adding $\beta_1 \bar{x} - \beta_1 \bar{x}$, which equals zero, to the right-hand side, and combining terms, we can write

$$y_i = \beta_0 + \beta_1 \bar{x} + \beta_1 x_i - \beta_1 \bar{x} + e_i$$
$$= (\beta_0 + \beta_1 \bar{x}) + \beta_1 (x_i - \bar{x}) + e_i$$

Define $\alpha = \beta_0 + \beta_1 \bar{x}$ (α does not depend on i), and we can rewrite the last equation in the equivalent form,

$$y_i = \alpha + \beta_1 (x_i - \bar{x}) + e_i \qquad i = 1, 2, \ldots, n \tag{1.8}$$

This is called the *deviations from the sample average* form for simple regression. The least squares estimates are

$$\hat{\alpha} = \bar{y} \qquad \hat{\beta}_1 \text{ as given by (1.7)} \tag{1.9}$$

Forbes' data. The four quantities *SXX*, *SXY*, $\bar{x}$, and $\bar{y}$ are needed for computing the least squares estimators. These are

$$\bar{x} = 202.95294118 \quad SXX = 530.78235294 \quad SXY = 475.29570589$$
$$\bar{y} = 139.60588235 \quad SYY = 427.76281177 \tag{1.10}$$

The quantity SYY, although not yet needed, is given for completeness. Also the number of digits given in each of these computations is excessive, since there are at most four significant digits in the original data (the logarithms of the pressures were rounded to the digits shown in Table 1.1 to enable the interested reader to reproduce the above calculations). However, since these are intermediate calculations, they should be done as accurately as possible, and rounding should be done only to final results. Using the calculations given, we find

$$\hat{\beta}_1 = \frac{SXY}{SXX} = 0.895$$
$$\hat{\beta}_0 = \bar{y} - \hat{\beta}_1 \bar{x} = -42.131.$$

The reason for multiplying the log(pressure) by 100 may now be evident. If this had not been done, then the estimate of β_1 would have been 100 times smaller, that is, 0.00895, and such small numbers often lead to mistakes. In the deviations from the average form of the simple regression model, the estimate of the slope is as given above, and the estimate of α is $\hat{\alpha} = \bar{y}$ = 139.606.

The estimated line, given by either of the equations

$$\hat{y} = -42.131 + 0.895x$$
$$= 139.606 + 0.895(x - 202.953)$$

has been drawn in Figure 1.1. As previously noted, the resemblance of this fitted line to the data appears to be excellent.

1.3   Estimating σ^2

Ideally, an estimate of σ^2 that does not depend on the appropriateness of the fitted model is desirable. Generally, such an estimate is obtainable only in data sets with several values of y recorded at each of several values of x or from prior information, as described in Sections 4.2 and 4.3. Lacking these special situations, the usual estimate of σ^2 is model dependent, as it is a function of the residual sum of squares, $RSS = \sum \hat{e}_i^2$.

Since σ^2 is essentially the average squared size of the e_i's, we should expect that its estimator, to be called $\hat{\sigma}^2$, is obtained by averaging RSS, the sum of observed squared errors. Under the assumption that the e_i's are uncorrelated random variables with zero means and common variance σ^2, one can show that an unbiased estimate of σ^2 is obtained by dividing RSS by its *degrees of freedom* (d.f.), where

residual d.f. = number of cases − number of parameters in model

For simple regression, residual d.f. $= n - 2$. The rule above will apply to multiple regression models as well. Thus the estimate of σ^2 is given by

$$\hat{\sigma}^2 = \frac{RSS}{n - 2} \tag{1.11}$$

This quantity is called the *residual mean square*. In general, any sum of squares divided by its degrees of freedom is called a mean square.

To compute $\hat{\sigma}^2$, it is left to the reader (Exercise 1.6) to show that

$$RSS = SYY - \frac{(SXY)^2}{SXX} = SYY - \hat{\beta}_1^2 SXX \tag{1.12}$$

For Forbes' data,

$$RSS = 427.76281177 - \frac{(475.29570589)^2}{530.78235294}$$

$$= 2.15332 \tag{1.13}$$

(or 2.153 to 4 digits) and

$$\hat{\sigma}^2 = \frac{2.15332}{17 - 2} = 0.14355$$

(or 0.144 to three digits—the more accurate figure will be needed later). The square root of this quantity, $\hat{\sigma} = \sqrt{0.144} = 0.379$ is often called the *standard error of regression*. It is in the same units as is the variable Y; for Forbes' data, the units are $100 \times \log(\text{pressure})$.

If any of the assumptions (independence, zero means, common variance) concerning the e_i are incorrect, then it is clear that $\hat{\sigma}^2$ may not be a useful estimate of scale or dispersion for the e_i's. If there are nonnegligible fixed (lack of fit) components in the e_i's then the residual mean square may seriously overestimate the variance, while nonindependence may lead to an estimate of σ^2 that is too big or too small. Finally, if the variance is not constant from case to case, then, unless additional information is available, a single estimate of variance may not be useful.

On the other hand, if, in addition to the assumptions made previously, the e_i's are drawn from a normal distribution, then the residual mean square will be distributed as a multiple of a chi-squared random variable with $n - 2$ degrees of freedom, or symbolically,

$$(n - 2) \frac{\hat{\sigma}^2}{\sigma^2} \sim \chi^2(n - 2)$$

This fact, proved in more advanced books on linear models, is used to obtain the distribution of test statistics, and also to make confidence statements concerning σ^2. In particular, this fact implies that

$$E(\hat{\sigma}^2) = \sigma^2$$

although normality is not required for unbiasedness.

1.4   Properties of least squares estimates

The least squares estimates depend on data only through the aggregate statistics given in Table 1.2. This is both an advantage, making computing easy, and a disadvantage, since any two data sets for which these are identical give the same fitted regression, even if a straight line model is appropriate for one but not the other (as in Example 5.1 in Chapter 5). Also, the estimates $\hat{\beta}_0$ and $\hat{\beta}_1$ are linear in $y_1, \ldots, y_n$. From (1.8), the fitted value at $x = \bar{x}$ is $\hat{y} = \bar{y} + \hat{\beta}(\bar{x} - \bar{x}) = \bar{y}$, so the fitted line must pass through the point $(\bar{x}, \bar{y})$, intuitively the center of the data. Finally, it is easy to show that $\sum \hat{e}_i = 0$, so negative and positive residuals balance out. Actually, $\sum \hat{e}_i = 0$ is a consequence of fitting an intercept term β_0, and models in which no intercept term appears will usually have $\sum \hat{e}_i \neq 0$.

Now, if the e_i's are random variables, then so are the estimates of β_0 and β_1, since these depend on the y_i's and hence on the e_i's. If all the e_i's have zero mean, $E(e_i) = 0$, $i = 1, \ldots, n$ and the model is correct, then, as shown in Appendix 1A.4, the least squares estimates are unbiased,

$$E(\hat{\beta}_0) = \beta_0$$
$$E(\hat{\beta}_1) = \beta_1$$

For the variance of the estimators, we now consider only the special case of $\text{var}(e_i) = \sigma^2$, $(i = 1, \ldots, n)$ and $\text{cov}(e_i, e_j) = 0$, $i \neq j$. Then, from Appendix 1A.4,

$$\text{var}(\hat{\beta}_1) = \sigma^2 \frac{1}{SXX}$$
$$\text{var}(\hat{\beta}_0) = \sigma^2 \left(\frac{1}{n} + \frac{\bar{x}^2}{SXX} \right) \tag{1.14}$$

In the somewhat simpler deviations from the sample average model, the variance of $\hat{\alpha}$ is given by

$$\text{var}(\hat{\alpha}) = \frac{\sigma^2}{n} \tag{1.15}$$

One important difference between the parameterization of the model (1.2) and that of (1.8) is the fact that the estimates $\hat{\beta}_0$ and $\hat{\beta}_1$ are correlated, as outlined in Appendix 1A.4,

$$\text{cov}(\hat{\beta}_0, \hat{\beta}_1) = -\sigma^2 \frac{\bar{x}}{SXX} \qquad (1.16)$$

but the estimates $\hat{\beta}_1$ and $\hat{\alpha}$ are uncorrelated, a fact that makes other computations, such as that of the variance of a prediction, relatively simple.

With the assumptions that the e_i are uncorrelated random variables with common variance, we can apply the Gauss-Markov theorem: under these conditions, the least squares estimates, which are linear functions of the y_i's, have the smallest possible variance. This means that, if one believes the assumptions, and is interested in using linear unbiased estimates, the least squares estimates are the ones to use.

If we add the final condition that the errors are normally distributed, then we can find the distributions of $\hat{\beta}_0$ and $\hat{\beta}_1$. Under the assumption that

$$e_i \sim \text{NID}(0, \sigma^2) \qquad i = 1, \ldots, n$$

then $\hat{\beta}_0$ and $\hat{\beta}_1$ are also normally distributed (since they are linear functions of the y_i's and hence of the e_i's) with variances and covariances given by (1.14) and (1.16). These results are used in obtaining estimates and confidence intervals.

Estimated variances. Estimates of $\text{var}(\hat{\beta}_0)$ and $\text{var}(\hat{\beta}_1)$ are obtained by substituting $\hat{\sigma}^2$ for σ^2 in (1.14). We use the symbol vâr( ) to mean estimated variances. Thus

$$\begin{aligned} \text{vâr}(\hat{\beta}_0) &= \hat{\sigma}^2 \left(\frac{1}{n} + \frac{\bar{x}^2}{SXX} \right) \\ \text{vâr}(\hat{\beta}_1) &= \hat{\sigma}^2 \frac{1}{SXX} \end{aligned} \qquad (1.17)$$

The square root of an estimated variance is called a *standard error*, for which we use the symbol se( ). The use of this notation is illustrated by

$$\text{se}(\hat{\beta}_1) = \sqrt{\text{vâr}(\hat{\beta}_1)}$$

1.5   Comparing models: the analysis of variance

The analysis of variance provides a convenient method of comparing the fit of two or more models to the same set of data. The methodology developed here is very useful in multiple regression, although all of the important principles can now be illustrated.

An elementary alternative to the simple regression model suggests fitting the equation

$$y_i = \beta_0 + e_i \qquad i = 1, 2, \ldots, n \qquad (1.18)$$

This model asserts that the y_i's depend on a single parameter β_0 plus random variation, but not on x_i. Fitting this model is equivalent to finding a line parallel to the x axis, as shown in Figure 1.4. The least squares line is $\hat{y} = \hat{\beta}_0$, where $\hat{\beta}_0$ is chosen to minimize $\sum (y_i - \hat{\beta}_0)^2$. It is easy to show that for this model

$$\hat{\beta}_0 = \bar{y} \qquad (1.19)$$

The residual sum of squares is

$$\sum (y_i - \hat{\beta}_0)^2 = \sum (y_i - \bar{y})^2 = SYY \qquad (1.20)$$

This residual sum of squares has $n - 1$ degrees of freedom (n cases minus 1 parameter in the model).

Next, consider the simple regression model obtained from (1.18) by adding a term that depends on x_i,

$$y_i = \beta_0 + \beta_1 x_i + e_i \qquad i = 1, 2, \ldots, n \qquad (1.21)$$

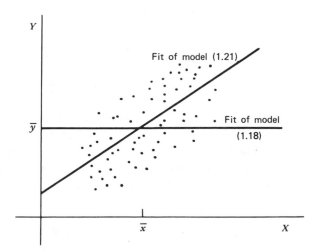

Figure 1.4   Two models compared by the analysis of variance.

Fitting this model is equivalent to finding the best line of arbitrary slope, as shown in Figure 1.4. The least squares estimates for this model are given by (1.7). As an important aside, we see that the estimates of β_0 under the two models are different, just as the meanings of the parameters in the two models are different. For (1.18), $\hat{\beta}_0$ is the average of the y's, but for (1.20), $\hat{\beta}_0$ is the average when $x_i = 0$.

For (1.21), the residual sum of squares, given in (1.12), is

$$RSS = SYY - \frac{(SXY)^2}{SXX} \tag{1.22}$$

As mentioned earlier, RSS has $n - 2$ degrees of freedom.

The difference between the sum of squares at (1.20) and that at (1.22) is the reduction in residual sum of squares due to enlarging the model from that at (1.18) to the simple regression model (1.21). This is the *sum of squares due to regression*, *SReg*, defined by

$$
\begin{aligned}
SSreg &= SYY - RSS \\
&= SYY - \left(SYY - \frac{(SXY)^2}{SXX} \right) \\
&= \frac{(SXY)^2}{SXX}
\end{aligned}
\tag{1.23}
$$

The degrees of freedom associated with $SSreg$ is the difference in degrees of freedom for the model (1.18), $n - 1$ and the d.f. for model (1.21), $n - 2$, so the d.f. for $SSreg$ is $(n - 1) - (n - 2) = 1$ for simple regression.

These results are often summarized in an analysis of variance table, abbreviated as ANOVA, given in Table 1.3. In the analysis of variance table, the column marked source refers to descriptive labels given to the sums of squares; in more complicated tables, there may be many sources, and the labels given are not unique. The degrees of freedom (d.f.) column gives the number of degrees of freedom associated with the named source. The next column gives the associated sum of squares. The mean square column is computed from the sum of squares column by dividing sums of squares by the corresponding degrees of freedom. The mean square on the residual line is just $\hat{\sigma}^2$, as already discussed.

Table 1.3   Analysis of variance

Source	Degrees of Freedom (d.f.)	Sum of squares (SS)	Mean square (MS)	F
Regression on X	1	$SSreg$	$SSreg/1$	$MSreg/MSE$
Residual for larger model	$n - 2$	RSS	$RSS/(n-2)$	
Total corrected sum of squares	$n - 1$	SYY		

The analysis of variance for Forbes' data is given in Table 1.4. For this table SYY is given at (1.10), RSS is given at (1.13), and $SSreg$ is computed by subtraction.

Table 1.4   Analysis of variance

Source	d.f.	SS	MS	F
Regression	1	425.610	425.610	2955
Residual	15	2.153	0.144	
Total	16	427.763		

It is important to note that the ANOVA is computed relative to a specific larger model given by (1.21) and a smaller model obtained from the full model by setting some parameters to zero. This was done at (1.18), which was obtained from (1.21) by setting $\beta_1 = 0$. The line in the ANOVA table corresponding to the total gives the residual sum of squares corresponding to the model with the fewest parameters. In the next chapter, the analysis of variance is applied to a sequence of models, but the reference to a fixed large model remains intact.

In practice, the ANOVA table is computed by finding SYY and $SSreg = SXY^2/SXX$, or by some equivalent formula from Table 1.2. RSS is computed by subtraction.

The F-test for regression. If the sum of squares for regression $SSreg$ is large, then the simple regression model $y_i = \beta_0 + \beta_1 x_i + e_i$ should be a significant improvement over the model given by (1.18), $y_i = \beta_0 + e_i$. This is equivalent to saying that the additional parameter in the simple regression model, β_1, is different from zero, or that Y is indeed related at X. To formalize this notion, we need to be able to judge how large is "large." This is done by comparing the regression mean square ($SSreg$ divided by its degree of freedom, which, for simple regression, is 1) to the residual mean square $\hat{\sigma}^2$. We call this ratio F:

$$F = \frac{(SYY - RSS)/1}{\hat{\sigma}^2} = \frac{SSreg/1}{\hat{\sigma}^2} \qquad (1.24)$$

Clearly F is a rescaled version of $SSreg = SYY - RSS$, with larger values of $SSreg$ resulting in larger values of F. Formally, we can consider testing the *null hypothesis* (NH) against the *alternative hypothesis* (AH)

$$\begin{aligned} \text{NH:} \quad & y_i = \beta_0 + e_i && i = 1, 2, \ldots, n \\ \text{AH:} \quad & y_i = \beta_0 + \beta_1 x_i + e_i && i = 1, 2, \ldots, n \end{aligned} \qquad (1.25)$$

If the e_i's are $NID(0, \sigma^2)$ then, under NH, (1.24) will follow an F distribution with degrees of freedom associated with the numerator and denominator of (1.24), 1 and $n - 2$ for simple regression. This is written $F \sim F(1, n - 2)$. Thus, percentage points of the F distribution can be used to assign significance levels, or p-values, to the computed F-test.

For Forbes' data, we compute

$$F = \frac{425.610}{0.144} = 2955$$

with (1, 15) degrees of freedom. From Table B at the end of the book, we see that the 0.01 point of $F(1, 15)$, written as $F(0.01; 1, 15) = 8.68$, so that the p-value for this test is (much) smaller than 0.01, providing very strong evidence against NH and in favor of AH.

Interpreting p-values. Under the appropriate assumptions, the p-value is the conditional probability of observing a value of the computed statistic (here, the value of F) as extreme or more extreme (here, as large or larger) than the observed value, given that the NH is true. A small p-value provides evidence against the NH. The observed p-value will depend on the sample size, the sampling plan, and on how far the correct AH is from the NH. Large (in absolute value) β_1's will generally lead to smaller p-values than would smaller values of β_1. Similarly, as sample size increases, p-values will generally get smaller since the power of the F-test increases with sample size, and the F-test will detect less extreme alternative hypotheses. Also, in regression situations, the p-value will depend on the sample range for the x's in the data; if the x's are obtained over a small range, the p-value will be larger than if the x's are sampled over a wider range.

Thus there is an important distinction between statistical significance, the observation of a sufficiently small p-value, and scientific significance, observing an effect of sufficient magnitude to be meaningful. Judgment of the latter will usually require examination of more than just the p-value.

1.6 The coefficient of determination, R^2

If both sides of (1.23) are divided by SYY, we get

$$\frac{SSreg}{SYY} = 1 - \frac{RSS}{SYY} \qquad (1.26)$$

The left-hand side of (1.26) is the proportion of variability explained by regression on X or equivalently by adding X to the model. The right-hand side consists of 1 minus the remaining unexplained variability. This concept of dividing up the total variability according to whether or not it is explained is of sufficient importance that a special name is given to it. We define R^2, the *coefficient of determination*, to be

$$R^2 = \frac{SSreg}{SYY} = 1 - \frac{RSS}{SYY} \qquad (1.27)$$

R^2 is easily computed from quantities that are available in the analysis of variance table. It is a scale-free one-number summary of the strength of the relationship between the x_i and the y_i in the data. It is a very popular statistic since it generalizes nicely to multiple regression, depends only on the sums of squares, and appears to be easy to interpret. For Forbes' data,

$$R^2 = \frac{SSreg}{SYY} = \frac{425.610}{427.763} = 0.995$$

and thus about 99.5% of the variability in the observed values of $100 \times \log(\text{pressure})$ is explained by boiling point.

Relationship to the correlation coefficient. By appealing to (1.27) and to Table 1.2, we can write

$$R^2 = \frac{SSreg}{SYY} = \frac{(SXY)^2}{(SXX)(SYY)} = r_{XY}^2$$

and thus R^2 is the same as the square of the sample correlation between X and Y.

1.7 Confidence intervals and tests

When the errors are $NID(0, \sigma^2)$, then parameter estimates, fitted values, and predictions will be normally distributed because all of these are linear combinations of the y_i's and hence of the e_i's. Consequently, confidence intervals and tests can be based on the t distribution. Suppose we let $t(\alpha, d)$ be the value of the t distribution with d degrees of freedom that cuts off $\alpha/2 \times 100\%$ in the lower tail and $\alpha/2 \times 100\%$ in the upper tail (e.g., the two-tailed $\alpha \times 100\%$ point). These values are given in Table A at the end of the book.

The intercept. The standard error of the intercept is $se(\hat{\beta}_0) = \hat{\sigma}(1/n + \bar{x}^2/SXX)^{1/2}$. Hence a $(1 - \alpha) \times 100\%$ confidence interval for the intercept is the set of points β_0 in the interval

$$\hat{\beta}_0 - t(\alpha, n - 2)se(\hat{\beta}_0) \leqslant \beta_0 \leqslant \hat{\beta}_0 + t(\alpha, n - 2)se(\hat{\beta}_0) \qquad (1.28)$$

For Forbes' data, $se(\hat{\beta}_0) = 0.0379(1/17 + (202.95)^2/530.78)^{1/2} = 3.339$. For a 90% confidence interval, $t(0.10, 15) = 1.75$, and the interval is

$$-42.131 - 1.75(3.339) \leqslant \beta_0 \leqslant -42.131 + 1.75(3.339)$$
$$-47.974 \leqslant \beta_0 \leqslant -36.288 \tag{1.29}$$

Ninety percent of such intervals will include the true value.

A hypothesis test of

$$\text{NH:} \quad \beta_0 = \beta_0^*, \quad \beta_1 \text{ arbitrary}$$
$$\text{AH:} \quad \beta_0 \neq \beta_0^*, \quad \beta_1 \text{ arbitrary}$$

is obtained by computing the t statistic

$$t = \frac{\hat{\beta}_0 - \beta_0^*}{se(\hat{\beta}_0)} \tag{1.30}$$

and referring this ratio to the t distribution with $n - 2$ degrees of freedom. For example, in Forbes' data consider testing

$$\text{NH:} \quad \beta_0 = -35, \quad \beta_1 \text{ arbitrary}$$
$$\text{AH:} \quad \beta_0 \neq -35, \quad \beta_1 \text{ arbitrary} \tag{1.31}$$

The statistic is

$$t = \frac{-42.121 - (-35)}{3.339} = 2.136 \tag{1.32}$$

which has a p-value near 0.05, $t(0.05, 15) = 2.13$, providing some evidence against NH. Of course, this hypothesis test for these data is not one that would occur to most investigators, and is used only as an illustration.

Slope. The standard error of $\hat{\beta}_1$ is $se(\hat{\beta}_1) = \hat{\sigma}/\sqrt{SXX} = 0.0164$. A 95% confidence interval for the slope is the set of β_1 such that

$$0.895 - 2.13(0.0164) < \beta_1 < 0.895 + 2.13(0.0164)$$
$$0.860 < \beta_1 < 0.930 \tag{1.33}$$

As with the intercept, hypothesis tests for the slope are probably not of interest in this example, as there is no obvious value for the parameter for a null hypothesis. The usual test is

$$\text{NH:} \quad \beta_1 = 0, \quad \beta_0 \text{ arbitrary}$$
$$\text{AH:} \quad \beta_1 \neq 0, \quad \beta_0 \text{ arbitrary} \tag{1.34}$$

For the data given, $t = (0.895 - 0)/0.0164 = 54.45$. Although clearly unnecessary here, since t is so large, we could compare this computed value to the t distribution with 15 d.f. The associated p-value is very small, but this is hardly surprising. If β_1 were actually near zero, Forbes probably would not have ever done the experiments, or if he had, he never would have published the results.

Compare the hypothesis (1.34) to (1.25). Both appear to be identical. In fact,

$$
t^2 = \left(\frac{\hat{\beta}_1}{\operatorname{se}(\hat{\beta}_1)} \right)^2 = \frac{\hat{\beta}_1^2}{\hat{\sigma}^2/SXX} = \frac{\hat{\beta}_1^2 SXX}{\hat{\sigma}^2} = F
$$

so the square of a t statistic with d degrees of freedom is equivalent to an F statistic with $(1, d)$ degrees of freedom.

Fitted values. If one were to do repeated sampling of units with the same value x, the fitted response for each of the units would be $\hat{y} = \hat{\beta}_0 + \hat{\beta}_1 x$. The standard error of the fitted value, written, $\operatorname{sefit}(\hat{y} \mid x)$ would be

$$
\operatorname{sefit}(\hat{y} \mid x) = \hat{\sigma} \left(\frac{1}{n} + \frac{(x - \bar{x})^2}{SXX} \right)^{1/2} \tag{1.35}
$$

Note that $\operatorname{sefit}(\hat{y} \mid x)$ increases as x moves away from $\bar{x}$. A confidence interval for the fitted value corresponding to x is the set of points y, such that

$$
\hat{y} - t(\alpha, n - 2)\operatorname{sefit}(\hat{y} \mid x) < y < \hat{y} + t(\alpha, n - 2)\operatorname{sefit}(\hat{y} \mid x) \tag{1.36}
$$

Prediction. If a new case is obtained with $X = \tilde{x}$, the predicted value of Y is $\tilde{y} = \hat{\beta}_0 + \hat{\beta}_1 \tilde{x} + \tilde{e}$, where $\tilde{e}$ is the error term that is attached to the new case under study. On the average, assuming that the fixed component to the error is negligible, $\tilde{e}$ will be zero, so the actual prediction is $\tilde{y} = \hat{\beta}_0 + \hat{\beta}_1 \tilde{x}$, but $\tilde{e}$ will inflate the variability of the prediction. The standard error of prediction, written $\operatorname{sepred}(\tilde{y} \mid \tilde{x})$, is

$$
\operatorname{sepred}(\tilde{y} \mid \tilde{x}) = \hat{\sigma} \left(1 + \frac{1}{n} + \frac{(\tilde{x} - \bar{x})^2}{SXX} \right)^{1/2} \tag{1.37}
$$

This standard error has two components, the part due to the new error, and the part due to uncertainty in estimating the β's. Even if we knew the β's exactly, we could not predict $\tilde{y}$ with perfect accuracy.

A confidence interval for a single prediction is obtained in the usual way, using multipliers from the appropriate t distribution. For a prediction of $100 \times \log(\text{pressure})$ at $\tilde{x} = 200$, the predicted value is $\tilde{y} = -42.13 +$

0.895(200) = 136.87, with standard error of prediction

$$0.379\left(1 + \frac{1}{n} + \frac{(200 - \bar{x})^2}{SXX}\right)^{1/2} = 0.393$$

Thus a 99% prediction interval at $\tilde{x} = 200$ is the set of $\tilde{y}$'s such that

$$136.87 - 2.95(0.393) \leqslant \tilde{y} \leqslant 136.87 + 2.95(0.393)$$

$$135.71 \leqslant \tilde{y} \leqslant 138.03$$

Figure 1.5 is a plot of the least squares regression line for Forbes' data along with two curves at $(\beta_0 + \beta_1\bar{x}) + t(.01;15)$ sepred $(\tilde{y}/\tilde{x})$ for x in the range from 180° to 220°. The vertical distance between the two curves for any x corresponds to a 99% prediction interval for $\tilde{y}$ given $\tilde{x}$. The interval is wider for $\tilde{x}$'s far from $\bar{x}$, and the curves bend outward (this may be hard to see in the figure because the variation about the regression line is so small).

A plot similar to Figure 1.5 can be obtained to represent a $(1 - \alpha) \times 100\%$ confidence interval for the fitted regression line. When this is done multipliers should not be obtained from the t distribution, since this will result in an overall confidence level different from $(1 - \alpha) \times 100\%$. The appropriate multiplier to use is $[2F(\alpha;2,n-2)]^{1/2}$. (In multiple regression the multiplier is $[p'F(\alpha;p',n-p')]^{1/2}$, where p' is the number of parameters in the model.) Thus, a 95% simultaneous confidence interval for the fitted regression line is, for each x, the set of all y such

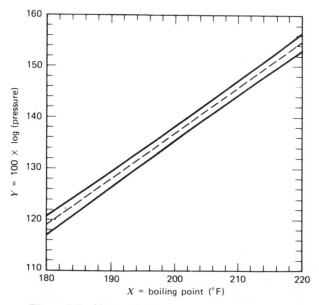

Figure 1.5   99% prediction intervals.

that

$$(\beta_0 + \beta_1 x) - [2F(.05;2,n-2)]^{1/2} \, \text{sefit} \, (\hat{y}/x) \leq y$$

$$\leq (\beta_0 + \beta_1 x) + [2F(.05;2,n-2)]^{1/2} \, \text{sefit} \, (\hat{y}/x) \quad (1.38)$$

The shape of the simultaneous confidence interval for the fitted line is similar to Figure 1.5, except that the curved lines are generally closer to the least squares line since for all x, sefit is smaller than sepred.

1.8 The residuals

The residuals, $\hat{e}_i = y_i - \hat{y}_i$, $i = 1, 2, \ldots, n$, provide information about assumptions about error terms and about the appropriateness of the model. Any complete data analysis requires examination of the residuals. Here we shall present only the barest outline of analysis of residuals, leaving detailed analysis to later chapters.

Plots of residuals versus other quantities are used to find failures of assumptions. The most important plot, especially in simple regression, is the plot of $\hat{e}_i$ versus the fitted values $\hat{y}_i$. Systematic features in this plot are of interest. Curvature might indicate that the fitted model is inappropriate, and suggest a transformation of the data. Residuals that seem to increase or decrease with $\hat{y}_i$ might indicate nonconstant residual variance. A few relatively large residuals may be indicative of outliers—cases for which the model is somehow inappropriate. On the other hand, if the plot of $\hat{e}_i$ versus $\hat{y}_i$ shows no systematic features, then we would have little reason to suspect that the fitted model was inappropriate for the data.

Forbes' data (conclusion and summary). The fitted values $\hat{y}_i$ and residuals $\hat{e}_i$ for Forbes' data are given in Table 1.5 and are plotted against each other in Figure 1.6. Notice that the residuals are generally small compared to the $\hat{y}_i$'s and that they do not suggest any distinct pattern in Figure 1.6. However, one residual, for case 12, is much larger than the others, as the others are typically less than 0.35 in absolute value, while that for case 12 is about 1.3. This *may* suggest that the assumptions concerning the errors are not correct. Either σ^2 may not be constant, or for case 12 the corresponding error may have a large fixed component. It is possible, for example, that Forbes misread or miscopied the results of his calculations for this case, and the numbers in the data do not correspond to the actual measurements. Forbes noted this possibility himself, by marking this pair of numbers in his paper as being "evidently in error" because of the large observed residual.

Table 1.5 Fitted values and residuals for Forbes' data

Case Number	x_i	y_i	$\hat{y}_i$	$\hat{e}_i$
1	194.50	131.79	132.04	− 0.25
2	194.30	131.79	131.86	− 0.07
3	197.90	135.02	135.08	− 0.06
4	198.40	135.55	135.53	0.02
5	199.40	136.46	136.42	0.04
6	199.90	136.83	136.87	− 0.04
7	200.90	137.82	137.77	0.05
8	201.10	138.00	137.95	0.05
9	201.40	138.06	138.22	− 0.16
10	201.30	138.05	138.13	− 0.08
11	203.60	140.04	140.19	− 0.15
12	204.60	142.44	141.08	1.36
13	209.50	145.47	145.47	0.00
14	208.60	144.34	144.66	− 0.32
15	210.70	146.30	146.54	− 0.24
16	211.90	147.54	147.62	− 0.08
17	212.20	147.80	147.89	− 0.09

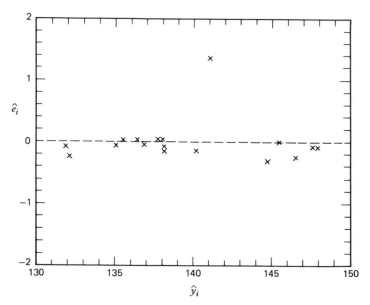

Figure 1.6 Residual plot for Forbes' data.

The problem of what to do given a suspect case must be faced. Lacking a formal procedure for examining such cases (a topic for later chapters), we proceed informally. Since we are concerned with the effects of case 12, we could refit the data, this time without case 12, and then examine the changes that occur in the estimates of parameters, fitted values, residual variance, and so on. This is summarized in Table 1.6, giving for each of the data sets (with 17 and 16 cases, respectively) estimates of parameters, their standard errors $\hat{\sigma}^2$, and the coefficient of determination R^2. As is clear from the table, for obtaining point estimates of the parameters case 12 is irrelevant, as the estimates are essentially identical with and without it. In other regression problems, deletion of a single case can change everything. However, the effect of case 12 on standard errors is more marked: if case 12 is deleted, standard errors are decreased by a factor of about 3.1 and variances are decreased by a factor of about $3.1^2 \cong 10$. Inclusion of this case therefore gives the appearance of less reliable results than would be suggested on the basis of the other 16 cases. Figure 1.7, a residual plot

Table 1.6 Summary for Forbes' data

Quantity	Value Using All Data	Value without Case 12
$\hat{\beta}_0$	− 42.131	− 41.302
$\hat{\beta}_1$	0.895	0.891
se($\hat{\beta}_0$)	3.339	1.000
se($\hat{\beta}_1$)	0.0164	0.00493
$\hat{\sigma}$	0.379	0.113
R^2	0.995	0.999 +

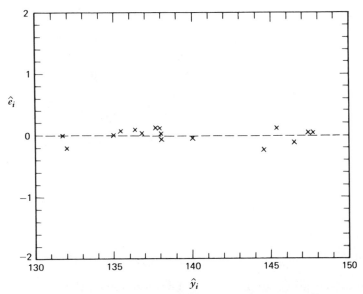

Figure 1.7 Residual plot for Forbes' data with case 12 deleted.

obtained when case 12 is deleted before computing, indicates no obvious failures in the model fit to the 16 remaining cases.

Two models have been fit, one using 16 cases and one using 17 cases, which lead to slightly different conclusions, although the results of the two analyses agree more than they disagree. On the basis of the data, there is no real way to choose between the two models, and we have no way of deciding which is the correct least squares analysis of the data. A good approach to this problem is to describe both (or in general all) plausible alternatives.

In summary, a straight line model of the data collected by Forbes can be used to describe the logarithm of atmospheric pressure as a function of boiling of water. For boiling points in the range 180 to 220°F, the data can be described by a straight line, with the equation given by $\hat{y} = -42.131 + 0.895x$, and, at least for any x in this range, this equation will provide a good prediction of $100 \times \log(\text{pressure})$. In the data, one case (number 12) is fit by the line much more poorly than are the other 16 points. Without this case, standard errors of estimates, predictions, etc., are only about one third of what they would be with it.

Even with the excellent fit of the straight line to the data, we cannot assume that the regression of $100 \times \log(\text{pressure})$ on boiling temperature is in fact a straight line; the conclusion should be restricted to the range of values in the observed data. The data provide no concrete information about the usefulness of the model beyond this range.

Problems

1.1 Height and weight data. These data give X = height in centimeters and Y = weight in kilograms for a sample of $n = 10$ eighteen-year-old girls. The data are taken from a larger study described in problem set 2.1.

X	Y
169.6	71.20
166.8	58.20
157.1	56.00
181.1	64.50
158.4	53.00
165.6	52.40
166.7	56.80
156.5	49.20
168.1	55.60
165.3	77.80

The following questions should be answered without the aid of a computer.

1.1.1. Draw a scatter plot of Y versus X. On the basis of this plot guess plausible values for β_0, β_1, and R^2 for the regression of Y on X in the simple linear regression model $y_i = \beta_0 + \beta_1 x_i + e_i$, $e_i \sim \text{NID}(0, \sigma^2)$, $i = 1$, $2, \ldots, 10$.

1.1.2. Show that $\bar{x} = 165.52$, $\bar{y} = 59.47$, $SXX = 472.076$, $SYY = 731.961$, and $SXY = 274.786$. Compute estimates of the slope and the intercept for the regression of Y on X. Draw the fitted line on your scatter plot. How many significant digits are there in $\hat{\beta}_0$ and $\hat{\beta}_1$?

1.1.3. Estimate σ^2 and find the estimated standard errors of $\hat{\beta}_0$ and $\hat{\beta}_1$. Also find the estimated covariance between $\hat{\beta}_0$ and $\hat{\beta}_1$. Compute the t-tests for the hypotheses that $\beta_0 = 0$ and that $\beta_1 = 0$, and find the appropriate p-values for these tests. (Use two-sided tests.)

1.1.4. Obtain the analysis of variance table and F-test for regression. Show numerically that $F = t^2$, where t was computed in 1.1.3 for testing $\beta_1 = 0$.

1.1.5. Obtain the residuals and the fitted values. Show numerically that the sum of the residuals is zero. Draw a graph of the residuals versus the fitted values. Are there any obvious outliers? Are there any clear patterns in the residuals?

1.1.6. Interpret the meaning of the parameters β_0 and β_1. What units are they measured in? What are the units of σ?

1.2   Hooker's data. In his paper on boiling points and temperatures, Forbes also presented data collected on the same two quantities by Dr. Joseph Hooker. Unlike Forbes, however, Hooker took his measurements in the Himalaya Mountains, generally at higher altitudes. The data below are

TEMP (°F)	PRES (in. Hg)	TEMP (°F)	PRES (in. Hg)
210.8	29.211	189.5	18.869
210.2	28.559	188.8	18.356
208.4	27.972	188.5	18.507
202.5	24.697	185.7	17.267
200.6	23.726	186.0	17.221
200.1	23.369	185.6	17.062
199.5	23.030	184.1	16.959
197.0	21.892	184.6	16.881
196.4	21.928	184.1	16.817
196.3	21.654	183.2	16.385
195.6	21.605	182.4	16.235
193.4	20.480	181.9	16.106
193.6	20.212	181.9	15.928
191.4	19.758	181.0	15.919
191.1	19.490	180.6	15.376
190.6	19.386		

an abstract of Hooker's data giving $n = 31$ pairs of measurements on TEMP = boiling point (degrees Fahrenheit) and PRES = corrected barometric pressure (inches of mercury). Use of packaged computer programs is encouraged, but not necessary, for this problem.

1.2.1. Draw the scatter plot of PRES versus TEMP. Would a straight line closely match the data? (A graph of the residuals from the regression PRES = $\beta_0 + \beta_1$TEMP + ERROR versus fitted values will be useful here.)

1.2.2. Draw a scatter plot of $100 \times \log$(PRES) versus TEMP, and compare to the plot in the last problem. Is this scatter plot more nearly described by a straight line?

1.2.3. Fit the simple regression model for $100 \times \log$(PRES) on TEMP; that is, fit the model $100 \times \log$(PRES) = $\beta_0 + \beta_1$TEMP + e and compute the relevant summary statistics (estimates of parameters, tests, analysis of variance table, R^2). Draw the fitted line onto the plot in problem 1.2.2. Obtain the residual plot (versus fitted values) and compare to that fit in 1.2.1.

1.2.4. Obtain 95% confidence intervals for β_0 and β_1.

1.2.5. Obtain 95% confidence interval for the fitted value at a boiling point of 185°F. Repeat for a boiling point of 212°F.

1.2.6. Obtain 90% prediction intervals for $100 \times \log$(PRES) for predictions at 185 and at 212°F. Obtain 95% prediction bands for all x in the range 180 to 220°F and compare to Figure 1.5.

1.2.7. Qualitatively compare the results of this analysis to the results in the test for Forbes' data. That is, compare the fitted lines, estimates of residual variability, prediction intervals, etc. What do you conclude? In Chapter 7 we will learn tests for comparing regression in different groups.

1.3.   Olympic records. The data below give the best foot race running times recorded in the modern Olympic Games up to 1976 (taken from the 1978 *Information Please Almanac*):

	Men			Women	
Distance (m)	Time (sec)	Year		Time (sec)	Year
100	9.9	1968		11.0	1968
200	19.8	1968		22.37	1976
400	43.8	1968		49.29	1972
800	103.5	1976		114.94	1976
1500	214.9	1968		241.40	1976
5000	804.8	1976			
10000	1658.4	1972			
42195	7795.0 (marathon)	1976			

A computer is not required for this problem.

1.3.1. Draw a scatter plot of time versus distance for all of the data, using a different symbol for men's times than for women's times (e.g., an $\times$ for men and a $\square$ for women). What do you learn from the graph?

1.3.2. Repeat 1.3.1, except replace time by speed, where speed = distance/time.

1.3.3. Perform relevant calculations that will help to summarize the relationship apparent in these data. You will probably want to do a separate analysis for men and for women.

1.3.4. Why are test statistics (t-tests and F-tests) not relevant in this problem?

1.4 Regression through the origin. Occasionally, a model in which the intercept is known a priori to be zero may be fit. This model is given by

$$y_i = \beta_1 x_i + e_i \qquad i = 1, 2, \ldots, n \qquad (1.39)$$

The residual sum of squares for this model, assuming the e_i are independent with common variance σ^2, is $RSS = \sum(y_i - \hat{\beta}_1 x_i)^2$.

1.4.1. Show that the least squares estimate of β_1 is given by $\hat{\beta}_1 = \sum x_i y_i / \sum x_i^2$. Show that $\hat{\beta}_1$ is unbiased and that $\text{var}(\hat{\beta}_1) = \sigma^2 / \sum x_i^2$. Find an expression for $\hat{\sigma}^2$. How many degrees of freedom does it have?

1.4.2. Derive the analysis of variance table with the larger model given by (1.21), but with the smaller model specified in (1.39). Show that the F-test derived from this table is numerically equivalent to the square of the t-test (1.30) with $\beta_0^* = 0$.

1.4.3. The data below give X = water content of snow on April 1 and Y = water yield from April to July (in inches) in the Snake River watershed in Wyoming for $n = 17$ years (1919 to 1935), from Wilm (1950).

X	Y	X	Y
23.1	10.5	37.9	22.8
32.8	16.7	30.5	14.1
31.8	18.2	25.1	12.9
32.0	17.0	12.4	8.8
30.4	16.3	35.1	17.4
24.0	10.5	31.5	14.9
39.5	23.1	21.1	10.5
24.2	12.4	27.6	16.1
52.5	24.9		

Fit a regression through the origin and find $\hat{\beta}_1$ and $\hat{\sigma}^2$. Obtain a 95% confidence interval for $\hat{\beta}_1$. Test the hypothesis that the intercept is zero.

1.4.4. Plot the residuals $(\hat{e}_i = y_i - \hat{\beta}_1 x_i)$ versus the fitted values $(\hat{y}_i = \hat{\beta}_1 x_i)$ and comment on the adequacy of the model. Note that $\sum \hat{e}_i \neq 0$ in regression through the origin.

1.5   Scale invariance.

1.5.1. In the simple regression model (1.2), suppose each x_i is replaced by cx_i, where $c \neq 0$ is a constant. How are $\hat{\beta}_0$, $\hat{\beta}_1$, $\hat{\sigma}^2$, R^2, and the t-test of NH: $\beta_1 = 0$ affected?

1.5.2. Suppose each y_i is replaced by dy_i, $d \neq 0$. Repeat 1.5.1.

1.6   Using Appendix 1A.3, verify equation (1.12).

1.7   Using Appendix 1A.4, verify equation (1.35).

1.8   Assuming model (1.2), verify that the sample correlation between the $\hat{e}_i$'s and the $\hat{y}_i$'s is zero. What is the sample correlation between the $\hat{e}_i$'s and the y_i's? In residual plotting suppose we plotted $\hat{e}_i$ versus y_i rather than $\hat{e}_i$ versus $\hat{y}_i$. Comment on the difference between these plots.

2

MULTIPLE REGRESSION

In multiple regression, several independent variables are used to model a single response variable. For each of the n cases observed, values for the response and for each of the independent variables are collected. If the response is called Y, and the independent variables are called X_1, $X_2, \ldots, X_p$ (p is the number of independent variables or predictors), then, the data will form the $n \times (p + 1)$ array:

Case Number	Values					
	Y	X_1	X_2	X_3	$\cdots$	X_p
1	y_1	x_{11}	x_{12}	x_{13}		x_{1p}
2	y_2	x_{21}	x_{22}	x_{23}		x_{2p}
$\vdots$	$\vdots$	$\vdots$	$\vdots$	$\vdots$	$\cdots$	$\vdots$
n	y_n	x_{n1}	x_{n2}	x_{n3}		x_{np}

For simple regression, $p = 1$. In the representation of the data, the value x_{ij} refers to the value for the jth variable on the ith case. The values for one case appear in one row; all the values for a variable appear in one column.

In multiple regression, an equation that expresses the response as a linear function of the p independent variables is estimated using the observed data. The model is specified by a linear equation,

$$Y = \beta_0 + \beta_1 X_1 + \beta_2 X_2 + \cdots + \beta_p X_p + e \qquad (2.1)$$

where, as in the previous chapter, the β's are unknown parameters, the e's are statistical errors, Y is the dependent variable or response, and X_1, $X_2, \ldots, X_p$ are the predictors. When $p = 2$, (2.1) gives the equation of a

31

two-dimensional plane in the three-dimensional (X_1, X_2, Y) space, as shown in Figure 2.1. Given data, x_{ij} collected on the X_j and y_i collected on Y, we rewrite (2.1) as

$$y_i = \beta_0 + \beta_1 x_{i1} + \beta_2 x_{i2} + \cdots + \beta_p x_{ip} + e_i \qquad i = 1, \ldots, n \qquad (2.2)$$

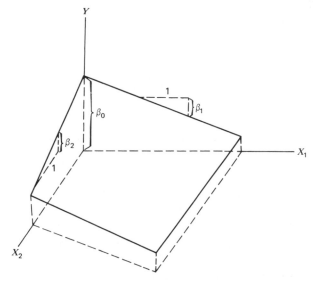

Figure 2.1   A linear regression surface with $p = 2$ regressors.

In this chapter, we will be concerned with estimating the β's and interpreting these estimates. Most of the results are given in terms of vectors and matrices, for which a brief introduction is given in Appendix 2A.1. With this notation, the results appear simple and elegant. Without it, one can get lost in a sea of subscripts.

Example 2.1   Fuel consumption

Six columns of Table 2.1 list values, for each of the 48 contiguous states, of the following quantities:

POP = 1971 population (in thousands).

TAX = 1972 motor fuel tax in cents per gallon.

NLIC = 1971 number of licensed drivers (in thousands).

INC = 1972 per capita personal income in dollars.

ROAD = 1971 length of Federal-aid primary highways, in miles.

FUELC = 1972 fuel consumption, in millions of gallons.

Table 2.1 Fuel consumption data

STATE	POP	X_1 TAX	NLIC	X_3 INC	X_4 ROAD	FUELC	X_2 DLIC	Y FUEL
1 ME	1029	9.00	540	3571	1976	557	0.525	541
2 NH	771	9.00	441	4092	1250	404	0.572	524
3 VT	462	9.00	268	3865	1586	259	0.580	561
4 MA	5787	7.50	3060	4870	2351	2396	0.529	414
5 RI	968	8.00	527	4399	431	397	0.544	410
6 CN	3082	10.00	1760	5342	1333	1408	0.571	457
7 NY	18366	8.00	8278	5319	11868	6312	0.451	344
8 NJ	7367	8.00	4074	5126	2138	3439	0.553	467
9 PA	11926	8.00	6312	4447	8577	5528	0.529	464
10 OH	10783	7.00	5948	4512	8507	5375	0.552	498
11 IN	5291	8.00	2804	4391	5939	3068	0.530	580
12 IL	11251	7.50	5903	5126	14186	5301	0.525	471
13 MI	9082	7.00	5213	4817	6930	4768	0.574	525
14 WI	4520	7.00	2465	4207	6580	2294	0.545	508
15 MN	3896	7.00	2368	4332	8159	2204	0.608	566
16 IA	2883	7.00	1689	4318	10340	1830	0.586	635
17 MO	4753	7.00	2719	4206	8508	2865	0.572	603
18 ND	632	7.00	341	3718	4725	451	0.540	714
19 SD	579	7.00	419	4716	5915	501	0.724	865
20 NE	1525	8.50	1033	4341	6010	976	0.677	640
21 KS	2258	7.00	1496	4593	7834	1466	0.663	649
22 DE	565	8.00	340	4983	602	305	0.602	540
23 MD	4056	9.00	2073	4897	2449	1883	0.511	464
24 VA	4764	9.00	2463	4258	4686	2604	0.517	547
25 WV	1781	8.50	982	4574	2619	819	0.551	460
26 NC	5214	9.00	2835	3721	4746	2953	0.544	566
27 SC	2665	8.00	1460	3448	5399	1537	0.548	577
28 GA	4720	7.50	2731	3846	9061	2979	0.579	631
29 FA	7259	8.00	4084	4188	5975	4169	0.563	574
30 KY	3299	9.00	1626	3601	4650	1761	0.493	534
31 TN	4031	7.00	2088	3640	6905	2301	0.518	571
32 AL	3510	7.00	1801	3333	6594	1946	0.513	554
33 MS	2263	8.00	1309	3063	6524	1306	0.578	577
34 AR	1978	7.50	1081	3357	4121	1242	0.547	628
35 LA	3720	8.00	1813	3528	3495	1812	0.487	487
36 OK	2634	6.58	1657	3802	7834	1695	0.629	644
37 TX	11649	5.00	6595	4045	17782	7451	0.566	640
38 MT	719	7.00	421	3897	6385	506	0.586	704
39 ID	756	8.50	501	3635	3274	490	0.663	648
40 WY	345	7.00	232	4345	3905	334	0.672	968
41 CO	2357	7.00	1475	4449	4639	1384	0.626	587
42 NM	1065	7.00	600	3656	3985	744	0.563	699
43 AZ	1945	7.00	1173	4300	3635	1230	0.603	632
44 UT	1126	7.00	572	3745	2611	666	0.508	591
45 NV	527	6.00	354	5215	2302	412	0.672	782
46 WN	3443	9.00	1966	4476	3942	1757	0.571	510
47 OR	2182	7.00	1360	4296	4083	1331	0.623	610
48 CA	20468	7.00	12130	5002	9794	10730	0.593	524

The data were collected by Christopher Bingham from the *American Almanac for 1974*, except for fuel consumption, which was given in the *1974 World Almanac and Book of Facts*. We shall use these data to study fuel consumption as a function of the other variables. Of particular interest will be the assessment of the relationship between TAX rate and FUEL consumption. In this study, other variables such as the number of fuel-consuming vehicles (autos, trucks, airplanes, and farm equipment) might be relevant, as might certain sociological indicators, such as a measure of the proportion of urban population, or indicators about the physical characteristics of the states, such as land area. For purposes of this chapter, we limit ourselves to the variables in Table 2.1 or transformations of them.

Before proceeding with the analysis, it is useful to consider further the nature of the measurements in the data. The variables NLIC and FUELC give measurements concerning the states as a whole, while INC has been scaled to a per person basis. Values of NLIC and FUELC will be relatively large in populous states and small in states with smaller populations, while INC will not depend directly on the population of the state. For example, large states like California and New York have fuel consumptions that are up to 30 times larger than the consumption in smaller states such as Vermont or Wyoming. We might therefore consider converting both FUELC and NLIC to rates by dividing them by POP. Then fuel consumption, income, and number of licensed drivers are all scaled to be on a per capita basis. The unit of analysis in the data would become the individual person within a state, rather than the state as a whole. In the remaining columns of Table 2.1, the data are given in scaled form. Further analysis in the chapter shall use these data only. The variables of interest are now*

X_1 = TAX, cents per gallon.

X_2 = DLIC = NLIC/POP = proportion of population with driver's licenses.

X_3 = INC, dollars.

X_4 = ROAD, miles.

Y = FUEL = 1000 × FUELC/POP = motor fuel consumption, gallons per person.

---

* A note on naming variables: In this book, the names $X_1, X_2, \ldots, X_p$ are the generic names of independent variables, while Y is the generic name for the dependent or response variable. In a particular problem, however, the variables used may well have other names, such as those given for the fuel consumption data. Most computer programs permit assigning up to 10-character descriptive names to variables, and that example is followed in this book.

The variable POP has been arbitrarily removed from the set of interesting variables to study because the response Y has been scaled to account for variations in POP.

The basic summary statistics, namely, sample means, standard deviations, and correlations, are given in Table 2.2. The sample correlation between DLIC and INC, for example, is 0.1517. Some of the numbers in this table are given in scientific notation; for example, the number $0.3290E + 06$ should be read as $0.3290 \times 10^6 = 329000$, and $0.3077E - 2 = 0.003077$.

Table 2.2 Basic summary statistics

Variable	N	Average	Variance	Standard Deviation	Minimum	Maximum
TAX	48	7.668	0.9040	0.9508	5.	10.
INC	48	4242.	$0.3290E+06$	573.6	3063.	5342.
DLIC	48	0.5703	$0.3077E-02$	$0.5547E-01$	0.451	0.724
ROAD	48	5565.	$0.1219E+08$	3492.	431.	$0.1778E+05$
FUEL	48	576.8	$0.1252E+05$	111.9	344.	968.

Matrix of sample correlations					
TAX	1.000				
DLIC	-0.2880	1.000			
INC	0.0127	0.1517	1.000		
ROAD	-0.5221	-0.0641	0.0501	1.000	
FUEL	-0.4513	0.6990	-0.2449	0.0190	1.000
	TAX	DLIC	INC	ROAD	FUEL

We will return to this example many times.

2.1 Adding a single independent variable to a simple regression model

Before turning to the general multiple regression model, we study the problem of adding one independent variable to a simple regression model. Specifically, in the fuel consumption data, we will consider adding TAX to a model for FUEL *after* DLIC has already been included. In the end, we will obtain an equation of the form

$$\widehat{FUEL} = \hat{\beta}_0 + \hat{\beta}_1 TAX + \hat{\beta}_2 DLIC \qquad (2.3)$$

(The hat notation is explained in Section 1.2.) First, we will fit the regression of FUEL on DLIC, and then TAX will be added to explain that part of the variation in FUEL not explained by DLIC.

To begin this procedure, scatter plots of FUEL versus DLIC and FUEL versus TAX are given in Figures 2.2 and 2.3. First consider Figure 2.2. This plot displays the relationship between FUEL and DLIC, ignoring all other variables. If we fit a simple regression model via least squares, we find

$$\widehat{FUEL} = 22.73 + 1409.8\,DLIC \tag{2.4}$$

with $R^2 = (0.6990)^2 = 0.4886$, indicating that 48.9% of the variability in FUEL is explained by DLIC, with a higher DLIC associated with higher FUEL consumption.

Similarly, the relationship between FUEL and TAX rate is summarized in Figure 2.3. The TAX rates occur in jumps of 0.5¢, and at some rates (e.g., 7.0¢) there is much more variability in FUEL than at others (9.0¢). Although this suggests that the simple regression model with constant residual variance may be inappropriate, we will fit this model as an intermediate step. The least squares line is

$$\widehat{FUEL} = 984.0 - 53.11\,TAX \tag{2.5}$$

with $R^2 = (-0.4513)^2 = 0.2037$, so TAX explains 20.4% of the variability in FUEL, *ignoring* DLIC, and each 1¢ increase in TAX rate, according to (2.5), corresponds to an estimated 53.11 gallon decrease in FUEL consumption.

Now suppose that we want to fit a model using both predictors to explain FUEL. The proportion of variability in FUEL explained by both DLIC and TAX will not equal 48.9% + 20.4% = 69.3% and may indeed be considerably less. The reason is that some of the variability in FUEL explained by DLIC may also be explained by TAX and the 69.3% may be double counting. Figure 2.4, a scatter plot of TAX versus DLIC, indicates that these two variables are related. The fitted regression equation is

$$\widehat{TAX} = 10.48 - 4.94\,DLIC \tag{2.6}$$

The interesting problem is to find the unique effect of adding TAX to a model that already includes DLIC. We are therefore concerned with modeling that part of FUEL that is not explained by DLIC by the part of TAX that is not explained by DLIC. Graphically, this requires examination of the scatter plot of the residuals from the regression of FUEL on DLIC versus the residuals from the regression of TAX on DLIC—in other words, the unexplained part of FUEL versus the unexplained part of TAX.

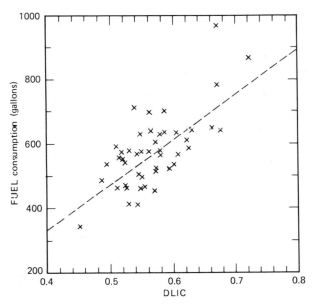

Figure 2.2   Scatter plot of FUEL versus DLIC.

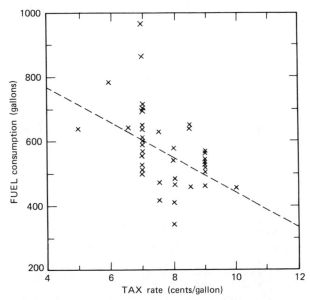

Figure 2.3   Scatter plot of FUEL versus TAX.

37

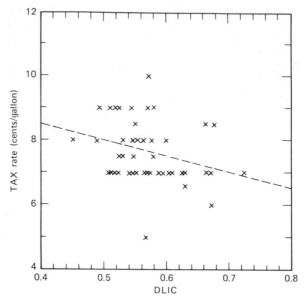

Figure 2.4 Scatter plot of TAX versus DLIC.

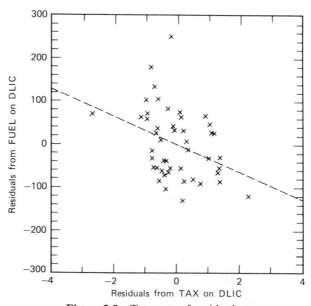

Figure 2.5 Two sets of residuals.

These residuals, obtained from (2.4) and (2.6), are plotted in Figure 2.5, which should be compared to Figure 2.3. The fitted slope in Figure 2.5 will give the relationship between FUEL and TAX adjusted for DLIC, while the fitted slope in Figure 2.3 gives the relationship between FUEL and TAX ignoring DLIC. The problem of unequal variances in Figure 2.3 has nearly disappeared after DLIC is fit, except perhaps for one point. If we fit a simple regression line to Figure 2.5, the fitted line has zero intercept, since the averages of the two variables plotted are zero, and estimated slope $\hat{\beta}_1$ $= -32.08$. It turns out that this is exactly the estimate of β_1 in (2.3), the multiple regression model.

Thus there are now estimates of β_1 from two models:

$$\begin{aligned} \hat{\beta}_1 &= -53.00 \quad \textit{ignoring DLIC} \\ \hat{\beta}_1 &= -32.08 \quad \textit{adjusted for DLIC} \end{aligned} \qquad (2.7)$$

While both of these indicate that higher TAX rates are associated with lower per capita FUEL consumption, the latter model, adjusting for DLIC, suggests that the magnitude of this effect is only about 60% as large as one might think if DLIC were ignored. In other regression problems, slope estimates for the same variable, but in different models, may be even more wildly different, changing signs, magnitude, and significance. This naturally complicates the interpretation of fitted models.

Just as $\hat{\beta}_1$ in (2.3) is an estimate of the effect of TAX adjusted for DLIC, so $\hat{\beta}_2$ is an estimate of the effect of DLIC adjusted for the effect of TAX. Thus $\hat{\beta}_2 = 1251.5$ is computed from the regression of the residuals from the regression of FUEL on TAX on the residuals from the regression of DLIC on TAX. In all multiple regression equations, the $\hat{\beta}_j$'s are adjusted for all other variables in the model.

To complete estimation of (2.3), we can compute $\hat{\beta}_0$ from the formula

$$\hat{\beta}_0 = \bar{y} - \sum_{j=1}^{p} \hat{\beta}_j \bar{x}_j = 109.0 \qquad (2.8)$$

where the sample averages are given in Table 2.2, and the β_j's are given above. The fitted equation for the two predictors is

$$\widehat{\text{FUEL}} = 109.0 - 32.08\,\text{TAX} + 1251.5\,\text{DLIC}$$

Partial correlation. In simple regression the relationship between Y and X_1 can be measured by the sample correlation r_{YX_1}. Similarly, the strength of the relationship between X_1 and Y adjusted for X_2 can be summarized by the correlation between the residuals for Y on X_2 and the residuals for X_1 on X_2, such as those plotted in Figure 2.5. This correlation is called a *partial correlation*, symbolically $r_{YX_1 | X_2}$, read as partial correlation between Y and X_1 adjusted for X_2.

Orthogonality. Two variables X_1 and X_2 are orthogonal if the regression of Y on X_1 adjusted for X_2 is *identical* to the regression of Y on X_1 ignoring X_2. This pleasant situation will occur if the sample correlation between X_1 and X_2, $r_{X_1 X_2}$, is *exactly* zero. When X_1 and X_2 are orthogonal, then the effect of each is unambiguous, and, consequently, experiments are often designed to have orthogonal variables.

2.2   Regression in matrix notation

Matrix notation will simplify most of the results used in multiple regression. In general, a vector or a matrix will be denoted by boldface letters like $\mathbf{X}$, $\mathbf{e}$, and $\boldsymbol{\beta}$. Elements of vectors and matrices are of the form x_{ij}, e_i, and β_j.

Let $\mathbf{Y}$ and $\mathbf{e}$ be $n \times 1$ vectors whose elements are given by the y_i's and the e_i's of (2.2), for example,

$$\mathbf{Y} = \begin{bmatrix} y_1 \\ y_2 \\ \vdots \\ y_n \end{bmatrix} \qquad \mathbf{e} = \begin{bmatrix} e_1 \\ e_2 \\ \vdots \\ e_n \end{bmatrix} \tag{2.9}$$

Also, define $\boldsymbol{\beta}$ to be the vector parameter of length $(p + 1) \times 1$, including the intercept β_0,

$$\boldsymbol{\beta} = \begin{bmatrix} \beta_0 \\ \beta_1 \\ \vdots \\ \beta_p \end{bmatrix} \tag{2.10}$$

Next, define $\mathbf{X}$ to be an $n \times (p + 1)$ matrix given by

$$\mathbf{X} = \begin{bmatrix} 1 & x_{11} & x_{12} \cdots x_{1p} \\ 1 & x_{21} & x_{22} \cdots x_{2p} \\ \vdots & \vdots & \vdots \\ 1 & x_{n1} & x_{n2} \cdots x_{np} \end{bmatrix} \tag{2.11}$$

As a matter of notational convenience, we will define p' to be the number of columns in X. For problems with an intercept term, $p' = p + 1$, and later we shall see that in problems without an intercept term, $p' = p$.

The matrix X gives all of the observed values of the independent variables, appended to a column of 1's as the leftmost column. X is called an incidence matrix, the data matrix, or the design matrix, depending on context. The ith row of X (also, the rows of e and Y) corresponds to values for the ith case in the data; the columns of X correspond to the different independent variables. We shall call the leftmost column of X, the column of 1's, the zeroth column, since this will be the column that corresponds to β_0. The next column, corresponding to the first independent variable X_1 and parameter β_1, will be the first column of X, and so on.

Using these quantities, the multiple regression equation (2.2) can be written in matrix terms as

$$Y = X\beta + e \tag{2.12}$$

The reader who is unfamiliar with matrices should use the definitions (2.9)–(2.11) to perform the indicated multiplications and additions in (2.12), and show that the results are, for the ith row, exactly the same as equation (2.2).

For the fuel consumption data, the first few and last few rows of the matrices X and Y are given by

$$
X = \begin{bmatrix}
1 & 9.0 & 0.525 & 3571 & 1976 \\
1 & 9.0 & 0.572 & 4092 & 1250 \\
1 & 9.0 & 0.580 & 3865 & 1586 \\
 & & \cdots & & \\
1 & 9.0 & 0.571 & 4476 & 3942 \\
1 & 7.0 & 0.623 & 4296 & 4083 \\
1 & 7.0 & 0.593 & 5002 & 9794
\end{bmatrix}
\qquad
Y = \begin{bmatrix}
541 \\
524 \\
561 \\
\cdots \\
510 \\
610 \\
524
\end{bmatrix}
$$

β is a parameter vector (of length $p' = 4 + 1 = 5$), and of course the values in β are unknown. The error vector e is unobserved.

Variance-covariance matrix of e. The error term is a vector of random variables called a random vector (as in Appendix 2A.2). The assumptions concerning the e_i's given in Chapter 1 are summarized in matrix form as $E(e) = 0$, var$(e) = \sigma^2 I_n$, where var(e) means the variance-covariance matrix of e, I_n is the $n \times n$ identity matrix, and 0 is an $n \times 1$ vector of zeroes. If we add the assumption that each e_i is normally distributed we will write

$$e \sim N(0, \sigma^2 I_n) \tag{2.13}$$

Least squares estimators. The least squares estimate $\hat{\beta}$ of β is chosen to minimize the residual sum of squares. Suppose we let x_i^T be the ith row of X, $i = 1, \ldots, n$ (using the convention that all vectors are columns, we affix a transpose to x_i to transform it to a row vector). Once $\hat{\beta}$ is obtained, the ith fitted value is $\hat{y}_i = x_i^T\hat{\beta}$ and the ith residual is $\hat{e}_i = y_i - \hat{y}_i = y_i - x_i^T\hat{\beta}$, and the residual sum of squares is $RSS = \sum \hat{e}_i^2$. Alternatively, we can write $\hat{Y} = X\hat{\beta}$ and $\hat{e} = Y - \hat{Y} = Y - X\hat{\beta}$, where $\hat{Y}$ is the vector of $\hat{y}_i$ and $\hat{e}$ is the vector of $\hat{e}_i$. The least squares estimator $\hat{\beta}$ is chosen to minimize

$$RSS = \hat{e}^T\hat{e} = (Y - \hat{Y})^T(Y - \hat{Y}) = (Y - X\hat{\beta})^T(Y - X\hat{\beta}) \qquad (2.14)$$

The least squares estimators can be found directly, from (2.14), in a matrix analog to the development of Appendix 1A.2. As an alternative to this, however, they can be found using an argument that leads to a numerically stable computational method. The basic idea is to transform the original least squares problem into an equivalent problem (via an orthogonal linear transformation) for which finding the least squares estimates is easy. The details of this can be found in Appendix 2A.3 and Problem 2.4. The least squares estimate $\hat{\beta}$ of β is

$$\hat{\beta} = (X^TX)^{-1}X^TY \qquad (2.15)$$

provided that $(X^TX)^{-1}$ exists. The estimator $\hat{\beta}$ depends only on (X^TX) and X^TY, which are the matrices of uncorrected sums of squares and cross products.

As in simple regression, an equivalent formula based on corrected sums of squares and cross products can be derived. To this end, let $\bar{x} = (\bar{x}_1, \bar{x}_2, \ldots, \bar{x}_p)^T$ be the $p \times 1$ vector of sample averages of the X's, and define $\mathcal{X}$ to be the $n \times p$ matrix of the original data with the averages subtracted off, so the (i, j)-th element of $\mathcal{X}$ is $x_{ij} - \bar{x}_j$. Similarly define $\mathcal{Y}$, an $n \times 1$ vector, to have ith element $y_i - \bar{y}$. Then the matrices of corrected cross products $\mathcal{X}^T\mathcal{X}$ and $\mathcal{X}^T\mathcal{Y}$, are

$$\mathcal{X}^T\mathcal{X} = \begin{bmatrix} \sum (x_{i1} - \bar{x}_1)^2 & \cdots & \sum (x_{i1} - \bar{x}_1)(x_{ip} - \bar{x}_p) \\ \vdots & & \vdots \\ \sum (x_{i1} - \bar{x}_1)(x_{ip} - \bar{x}_p) & \cdots & \sum (x_{ip} - \bar{x}_p)^2 \end{bmatrix} \qquad (2.16)$$

$$\mathcal{X}^T\mathcal{Y} = \begin{bmatrix} \sum (x_{i1} - x_1)(y_i - \bar{y}) \\ \vdots \\ \sum (x_{ip} - \bar{x}_p)(y_i - \bar{y}) \end{bmatrix} \qquad (2.17)$$

Generally, these matrices are presented as a single $(p + 1) \times (p + 1)$ matrix $\mathbf{T}$, where

$$\mathbf{T} = \begin{pmatrix} \mathcal{X}^T\mathcal{X} & \mathcal{X}^T\mathcal{y} \\ \mathcal{y}^T\mathcal{X} & \mathcal{y}^T\mathcal{y} \end{pmatrix} \begin{matrix} p \text{ rows} \\ 1 \text{ rows} \end{matrix} \qquad (2.18)$$

p columns   1 columns

Typically, computer programs are designed to print a matrix $\mathbf{S}$ of sample covariances and variances, which is obtained from $\mathbf{T}$ by dividing each element by $(n - 1)$, $\mathbf{S} = (n - 1)^{-1}\mathbf{T}$. Also, the sample correlation matrix given in Table 2.2 is obtained from $\mathbf{S}$ by the rule $r_{ij} = s_{ij}/\sqrt{s_{ii}s_{jj}}$. Since $\mathbf{S}$ is symmetric, only the lower triangular part is usually printed.

Suppose we let

$$\hat{\boldsymbol{\beta}}^* = (\mathcal{X}^T\mathcal{X})^{-1}\mathcal{X}^T\mathcal{y}$$

Then,

$$\hat{\beta}_0 = \bar{y} - \hat{\boldsymbol{\beta}}^{*T}\bar{\mathbf{x}} \qquad (2.19)$$

$$\hat{\boldsymbol{\beta}} = \begin{pmatrix} \hat{\beta}_0 \\ \hat{\boldsymbol{\beta}}^* \end{pmatrix}$$

Properties of the estimates. Properties of the least squares estimates are derived in Appendix 2A.3 and are only summarized here. Assuming $E(\mathbf{e}) = \mathbf{0}$ and $\mathrm{var}(\mathbf{e}) = \sigma^2\mathbf{I}_n$, then $\hat{\boldsymbol{\beta}}$ is unbiased $(E(\hat{\boldsymbol{\beta}}) = \boldsymbol{\beta})$, and

$$\mathrm{var}(\hat{\boldsymbol{\beta}}) = \sigma^2(\mathbf{X}^T\mathbf{X})^{-1} \qquad (2.20)$$

Also, σ^2 is estimated, according to the rule in Section 1.3, by

$$\hat{\sigma}^2 = \frac{RSS}{n - p'} \qquad (2.21)$$

RSS can be shown, by substituting $\hat{\boldsymbol{\beta}}$ into (2.14) and simplifying, to be any of the following equivalent expressions:

$$RSS = \mathbf{Y}^T\mathbf{Y} - \hat{\boldsymbol{\beta}}^T(\mathbf{X}^T\mathbf{X})\hat{\boldsymbol{\beta}}$$

$$= \mathbf{Y}^T\mathbf{Y} - \hat{\boldsymbol{\beta}}^T\mathbf{X}^T\mathbf{Y}$$

$$= \mathcal{y}^T\mathcal{y} - \hat{\boldsymbol{\beta}}^{*T}(\mathcal{X}^T\mathcal{X})\hat{\boldsymbol{\beta}}^*$$

$$= \mathcal{y}^T\mathcal{y} - \hat{\boldsymbol{\beta}}^T(\mathbf{X}^T\mathbf{X})\hat{\boldsymbol{\beta}} + n\bar{y}^2 \qquad (2.22)$$

As in Section 1.3, if $\mathbf{e}$ is normally distributed, then $\hat{\sigma}^2/\sigma^2$ has a $\chi^2(n - p')$ distribution.

By substituting (2.21) into (2.20), we find the estimated variance of

$\hat{\beta}, \widehat{\text{var}}(\hat{\beta})$, to be

$$\widehat{\text{var}}(\hat{\beta}) = \hat{\sigma}^2(\mathbf{X}^T\mathbf{X})^{-1} \tag{2.23}$$

Simple regression in matrix terms. For simple regression, $\mathbf{X}$ and $\mathbf{Y}$ are given by

$$\mathbf{X} = \begin{bmatrix} 1 & x_1 \\ 1 & x_2 \\ \vdots & \\ 1 & x_n \end{bmatrix} \qquad \mathbf{Y} = \begin{bmatrix} y_1 \\ y_2 \\ \vdots \\ y_n \end{bmatrix}$$

and thus

$$(\mathbf{X}^T\mathbf{X}) = \begin{bmatrix} n & \sum x_i \\ \sum x_i & \sum x_i^2 \end{bmatrix} \qquad \mathbf{X}^T\mathbf{Y} = \begin{bmatrix} \sum y_i \\ \sum x_i y_i \end{bmatrix} \tag{2.24}$$

$(\mathbf{X}^T\mathbf{X})^{-1}$ can be shown to be

$$(\mathbf{X}^T\mathbf{X})^{-1} = \frac{1}{SXX}\begin{pmatrix} \sum x_i^2/n & -\bar{x} \\ -\bar{x} & 1 \end{pmatrix} \tag{2.25}$$

so that

$$\hat{\beta} = \begin{pmatrix} \hat{\beta}_0 \\ \hat{\beta}_1 \end{pmatrix} = (\mathbf{X}^T\mathbf{X})^{-1}\mathbf{X}^T\mathbf{Y} = \frac{1}{SXX}\begin{pmatrix} \sum x_i^2/n & -\bar{x} \\ -\bar{x} & 1 \end{pmatrix}\begin{bmatrix} \sum y_i \\ \sum x_i y_i \end{bmatrix}$$

$$= \begin{bmatrix} \bar{y} - \hat{\beta}_1\bar{x} \\ \dfrac{SXY}{SXX} \end{bmatrix}$$

as found previously. Also, since $\sum x_i^2/(n)SXX = 1/n + \bar{x}^2/SXX$, then the variances and covariances for $\hat{\beta}_0$ and $\hat{\beta}_1$ found in the previous chapter are identical to those given by $\sigma^2(\mathbf{X}^T\mathbf{X})^{-1}$.

In the deviations from the sample average form, the results are simpler, since

$$\mathscr{X}^T\mathscr{X} = SXX \qquad \mathscr{X}^T\mathscr{y} = SXY$$

and

$$\hat{\beta}_1 = (\mathscr{X}^T\mathscr{X})^{-1}\mathscr{X}^T\mathscr{y} = \frac{SXY}{SXX}$$

$$\hat{\beta}_0 = \bar{y} - \hat{\beta}\bar{x}$$

Fuel consumption data (continued). We shall now fit the model with all $p = 4$ predictors to the fuel consumption data. We will write FUEL on

Table 2.3 $(X^T X)^{-1}$ for fuel consumption data

	Intercept	TAX	DLIC	INC	ROAD
Intercept	7.8301941	-0.42651325	-6.1107646	$-.14950904E-03$	$-.75349217E-04$
TAX	-0.42651325	$0.38263554E-01$	0.22158115	$-.59137167E-05$	$.57148375E-05$
DLIC	-6.1107646	0.22158115	8.4108914	$-.14500434E-03$	$.41268962E-04$
INC	$-0.14950904E-03$	$-0.59137167E-05$	$-.14500434E-03$	$.67459995E-07$	$-.15445116E-08$
ROAD	$-0.75349217E-04$	$0.57148375E-05$	$.41268962E-04$	$-.15445116E-08$	$.26126406E-08$

TAX DLIC ROAD INC to mean "fit the model $\widehat{FUEL} = \hat{\beta}_0 + \hat{\beta}_1$ TAX $+ \hat{\beta}_2$ DLIC $+ \hat{\beta}_3$ ROAD $+ \hat{\beta}_4$ INC." For these data and this model, $(\mathbf{X}^T\mathbf{X})^{-1}$ is given in Table 2.3 and $\mathbf{X}^T\mathbf{Y}$ and $\mathbf{Y}^T\mathbf{Y}$ are

$$\mathbf{X}^T\mathbf{Y} = \begin{bmatrix} 27685.000000 \\ 210041.52000 \\ 15993.565000 \\ 116696535.00 \\ 154428181.00 \end{bmatrix} \qquad \mathbf{Y}^T\mathbf{Y} = (1655627.0)$$

It can be shown that $(\mathcal{X}^T\mathcal{X})^{-1}$ is the lower right $p \times p$ submatrix of $(\mathbf{X}^T\mathbf{X})^{-1}$. Using the formulas in this chapter, the estimate $\hat{\beta}$ is computed as

$$\hat{\beta} = \begin{bmatrix} \hat{\beta}_0 \\ \hat{\beta}_1 \\ \hat{\beta}_2 \\ \hat{\beta}_3 \\ \hat{\beta}_4 \end{bmatrix} = \begin{bmatrix} 377.3 \\ -34.79 \\ 1336. \\ -0.06659 \\ -0.002426 \end{bmatrix}$$

The residual sum of squares is

$$RSS = \mathbf{Y}^T\mathbf{Y} - \hat{\beta}^T(\mathbf{X}^T\mathbf{X})\hat{\beta}$$

$$= 588,366 - 399,316 = 189,050$$

The residual mean square is

$$\hat{\sigma}^2 = \frac{RSS}{n - p'} = \frac{189,050}{48 - (4+1)} = 4396.5 \ (43 \text{ d.f.})$$

The standard errors and estimated covariances of the $\hat{\beta}_j$'s are found from $\hat{\sigma}^2$ and $(\mathbf{X}^T\mathbf{X})^{-1}$. For example,

$$se(\hat{\beta}_0) = \hat{\sigma}\sqrt{7.83019} = 185.54$$

$$se(\hat{\beta}_2) = \hat{\sigma}\sqrt{8.41089} = 192.30$$

$$\widehat{cov}(\hat{\beta}_1, \hat{\beta}_2) = \hat{\sigma}^2(0.22158) = 14.692$$

In most computer programs, the usual output obtained is somewhat less than that given here. The results in Table 2.4 are more typical of what might be expected. The first column gives the labels of the independent variables. The second column gives the corresponding estimated $\hat{\beta}_j$'s. The third column gives $\hat{\sigma}$ times the square root of the appropriate diagonal entry of $(\mathbf{X}^T\mathbf{X})^{-1}$. The last column ($t$-value) is the ratio $\hat{\beta}_j/se(\hat{\beta}_j)$, to be discussed shortly. Various other summary statistics, such as the number of d.f. for error, $\hat{\sigma}^2$ and/or $\hat{\sigma}$, and R^2 are generally also reported. Sometimes an excessive number of digits is printed, as is the case here. In these data, rounding results to at most four significant digits is appropriate.

Table 2.4   Computer program regression summary for the regression of FUEL on TAX INC ROAD DLIC

Variable	Estimate	Standard Error	t-Value
BO	377.2911	185.5412	2.03
TAX	-34.79015	12.9702	-2.68
DLIC	1336.449	192.2981	6.95
INC	$-0.6658875E-01$	$0.1722175E-01$	-3.87
ROAD	$-0.2425889E-02$	$0.3389174E-02$	-0.72

$\hat{\sigma}^2 = 4396.511$, d.f. $= 43$, $R^2 = 0.6787$.

Fitted values and predictions. The fitted value corresponding to a $p' \times 1$ vector $\mathbf{x}$ (with a 1 in the first element for the intercept) is given by $\hat{y} = \mathbf{x}^T\hat{\boldsymbol{\beta}}$. The standard error of this fitted value, $\text{sefit}(\hat{y}\,|\,\mathbf{x})$, using Appendix 2A.2, is

$$\text{sefit}(\hat{y}\,|\,\mathbf{x}) = \hat{\sigma}\sqrt{\mathbf{x}^T(\mathbf{X}^T\mathbf{X})^{-1}\mathbf{x}} \qquad (2.26)$$

A prediction at $\mathbf{x}$ is given by $\tilde{y} = \mathbf{x}^T\hat{\boldsymbol{\beta}} + \tilde{e}$, where we again include the $\tilde{e}$ to remind us that a prediction includes a random error $\tilde{e}$ that contributes to the standard error of prediction,

$$\text{sepred}(\tilde{y}\,|\,\mathbf{x}) = \hat{\sigma}\sqrt{1 + \mathbf{x}^T(\mathbf{X}^T\mathbf{X})^{-1}\mathbf{x}} \qquad (2.27)$$

2.3   The analysis of variance

For multiple regression, the analysis of variance is a very rich technique that is used to divide variability and to compare models that include different sets of variables. In the overall analysis of variance, the full model

$$\mathbf{Y} = \mathbf{X}\boldsymbol{\beta} + \mathbf{e} \qquad (2.28)$$

is compared to the model with no X variables,

$$\mathbf{Y} = \mathbf{1}\beta_0 + \mathbf{e} \qquad (2.29)$$

where $\mathbf{1}$ is a $n \times 1$ vector of ones. These correspond to (1.21) and (1.18), respectively. Thus, for model (2.29) $\hat{\beta}_0 = \bar{y}$ and the residual sum of squares is SYY. For model (2.28), on the other hand, the estimate of $\boldsymbol{\beta}$ is given in (2.15) and RSS is given in (2.22). Clearly, we must have $RSS < SYY$, and the difference between these two,

$$SSreg = SYY - RSS \qquad (2.30)$$

corresponds to the sum of squares in Y explained by the larger model that is not explained by the smaller model. The number of degrees of freedom associated with $SSreg$ is equal to the number of d.f. in SYY ($= n - 1$) minus the number of d.f. in RSS ($= n - p'$) or $(n - 1) - [n - (p + 1)] = p$.
These results are summarized in the analysis of variance table:

Source	d.f.	SS	MS
		Analysis of Variance (Overall)	
Regression on $X_1, \ldots, X_p$	p	$SSreg$	$SSreg/p$
Residual	$n - p - 1$	RSS	$RSS/(n - p - 1) = \hat{\sigma}_e^2$
Total	$n - 1$	SYY	

We can judge the importance of the regression on the X's by determining if $SSreg$ is sufficiently large by comparing the ratio of the mean square for regression to $\hat{\sigma}^2$ to the $F(p, n - p')$ distribution. If the computed F exceeds a convenient critical value, then we would judge that knowledge of the X's provides a significantly better model than does no knowledge of them. The ratio computed will have an exact F distribution if the errors are $NID(0, \sigma^2)$. The hypothesis tested by this F-test is

$$\text{NH:} \quad \text{model (2.29) applies,} \quad \beta^* = 0$$
$$\text{AH:} \quad \text{model (2.28) applies,} \quad \beta^* \neq 0$$

The coefficient of determination. As with simple regression the ratio

$$R^2 = \frac{SYY - RSS}{SYY} = \frac{SSreg}{SYY} \qquad (2.31)$$

gives the proportion of variability in Y explained by regression on the X. In addition, one can show that the value R^2 is equal to the *multiple correlation coefficient* between Y and the X's: it is the square of the maximum correlation between Y and any linear function of the X's.

Fuel consumption data. The overall analysis of variance table is given by

Source	d.f.	SS	MS	F
Regression	4	399,316	99,829	22.7
Residual	43	189,050	$4397 = \hat{\sigma}^2$	
Total	47	588,366		

Since $F = 22.7$ exceeds $F(0.01; 4, 43) = 3.82$ by a large margin, one would be led to suspect that at least some of the X's are in fact related to fuel consumption. The value of $R^2 = 399316/588366 = 0.68$, indicating that about 68% of the observed variability in the response is modeled by the X's. Without experience in problems like this one, it is not easy to decide if 68% is a lot or a little.

Incidentally, this example points out that the computation of the overall F statistic is not always interesting. Often it is known a priori that the variables are related so very large values of the test statistic are expected. Of more interest is examination of other hypotheses concerning some of the variables.

Hypotheses concerning one of the independent variables. In many problems, obtaining information on the usefulness of one of the predictors of the response is of interest. Can we do as well modeling fuel consumption from, for example, just DLIC, ROAD, and INC as we do from all four variables? This question can be rephrased in a more suggestive manner: if DLIC, ROAD, and INC are known, will knowledge of TAX represent a significant improvement? The following procedure can be used: fit the model that excludes TAX, and obtain the residual sum of squares for that model. Then, fit a second model including TAX, and get the residual sum of squares for this model. Subtracting the residual sums of squares for the larger model from the residual sum of squares for the smaller model will give the sum of squares for regression on TAX after adjusting for the variables that are already included in the model (e.g., DLIC, ROAD, and INC). This computation can be done exactly as outlined. Begin by performing the regression of Y on ROAD, DLIC, and INC. The residual sum of squares for this model is 220,682 (and $\hat{\sigma}^2 = 5015$). The residual sum of squares for the full model has already been given as 189,050 (and $\hat{\sigma}^2 = 4397$). The sum of squares for regression on TAX after the others is $220,682 - 189,050 = 31,632$ (and the estimated $\hat{\sigma}^2$ is reduced by about 7%). This can be summarized in the following analysis of variance table:

Source	d.f.	SS	MS	F
Regression on ROAD, INC, DLIC	3	367,684	122,561	
TAX after others	1	31,632	31,632	7.18
Residual	43	189,050	4,397	

The SS for regression on ROAD, INC, DLIC is found by subtracting the SSreg(ROAD, INC, DLIC, TAX) − SSreg(TAX after others) = 399,316 − 31,632 = 367,684. The ratio of mean squares $F = 31,632/4,397 = 7.18$ is the statistic used to test the usefulness of TAX after the other variables are already included in the model; it says nothing about the usefulness of the other variables. It is compared to the F distribution with 1 and $n - p' = 43$ degrees of freedom. In the example, TAX appears to be a significant predictor after adjusting for the others, since $F(0.01; 1, 43) = 7.25$, giving a p-value near 0.01. We call this a *partial* F-test. Specifically, this F tests the hypothesis

$$\text{NH:} \quad \beta_1 = 0; \quad \beta_0, \beta_2, \beta_3, \beta_4 \text{ arbitrary}$$
$$\text{AH:} \quad \beta_1 \neq 0; \quad \beta_0, \beta_2, \beta_3, \beta_4 \text{ arbitrary} \tag{2.32}$$

Relationship to the t statistic. Another reasonable procedure for testing the importance of TAX is simply to compare the estimate of the coefficient divided by its standard error to the t distribution with 43 degrees of freedom. It can be shown that the square of the t ratio is the same number as the F ratio just computed, so these two procedures are identical. Therefore, the t statistic tests hypotheses concerning the importance of variables adjusted for all the other variables in the model (*not* ignoring them).

For example, the t statistic for TAX is, from Table 2.4,

$$t = \frac{-34.79015}{12.97020} = -2.68$$

which would be compared to $t(43)$ to find critical values. The hypothesis tested by this statistic is given in (2.32). Also, one finds

$$t^2 = (-2.68)^2 = 7.18$$

numerically identical to the value obtained for the F-test for this hypothesis. A t-test that any of the β_j's has a specific value (given that all other β's are arbitrary) can be carried out as in Section 1.7.

Other tests of hypotheses. We have obtained a test of the hypothesis concerning the effects of TAX adjusted for all the variables in the problem. Equally well, we could obtain tests for the effect of TAX adjusting for *some* of the other variables, or for none of the other variables. In general, these tests will not be equivalent: a variable can be judged to be a useful predictor ignoring other variables, but judged as useless when adjusted for them. Furthermore, a predictor that is useless by itself may become important when considered in concert with the other variables. The outcome of these tests depends on the relationship between the X's as reflected in $\mathbf{X}^T\mathbf{X}$, or usually more clearly in the sample correlations. Therefore, a problem of order of fitting the various X's is apparent in multiple regression.

Sequential analysis of variance tables. By separating TAX from the other three independent variables, *SSreg* is divided into two pieces, one for fitting the first three variables and one for fitting TAX after the other three. This subdivision can be continued by dividing the sum of squares for regression into pieces for each variable. Unless all the independent variables are orthogonal, this breakdown is not unique. For example, we could first fit DLIC, then TAX adjusted for DLIC, then INC adjusted for DLIC and TAX, and finally ROAD adjusted for the other three. The resulting table is given in Table 2.5A. Alternatively, we could fit in the order ROAD, INC, DLIC, and then TAX as in Table 2.5B. As can be seen, the resulting associated sums of squares are quite different: fitting ROAD first, for example, has a sum of squares of only 213.3, but adjusted for the other three has a sum of squares of 2252, larger by a factor of 10, but still not very large compared to $\hat{\sigma}^2$. The reason is that several of the variables explain the same sum of squares.

Table 2.5   Two analyses of variance tables, different orders

	Source	d.f.	SS	MS
	A. First Analysis			
First	DLIC	1	$2.874E+05$	$2.874E+05$
Then	TAX	1	$4.008E+04$	$4.008E+04$
Then	INC	1	$6.953E+04$	$6.953E+04$
Then	ROAD	1	2252.	2252.
	Residual	43	$1.891E+05$	4397.
	B. Second Analysis			
First	ROAD	1	213.3	213.3
Then	INC	1	$3.564E+04$	$3.564E+04$
Then	DLIC	1	$3.318E+05$	$3.318E+05$
Then	TAX	1	$3.163E+04$	$3.163E+04$
	Residual	43	$1.890E+05$	4397.

2.4   Regression through the origin

Some regression problems may require fitting models through the origin by setting $\beta_0 = 0$. This is done in matrix notation by defining $\mathbf{X}$ to be an $n \times p$ matrix, *without* the column of ones, and $\boldsymbol{\beta}$ to be a $p \times 1$ matrix *without* β_0. Then, with these modifications, model (2.12) is valid, and all the matrix results following (2.12) apply, if $p' = p$ rather than $p + 1$.

Notational convention. Since the same results apply to regression through the origin, let $p' = p + 1$ if the intercept is in the model and $p' = p$ if regression is through the origin, and view $\mathbf{X}$ as an $n \times p'$ matrix and $\boldsymbol{\beta}$ as a $p' \times 1$ vector.

Problems

2.1   Berkeley Guidance Study. The data for this example are excerpted from the Berkeley Guidance Study, a longitudinal monitoring of boys and girls born in Berkeley, California between January 1928 and June 1929. The variables included in the data are:

X_1 = WT2 = weight at age 2 (kg).

X_2 = HT2 = height at age 2 (cm).

X_3 = WT9 = weight at age 9.

X_4 = HT9 = height at age 9.

X_5 = LG9 = leg circumference (cm).

X_6 = ST9 = a composite measure of strength at age 9 (high values = stronger).

X_7 = WT18 = weight at age 18.

Table 2.6   Berkeley Guidance Study: boys

Identi-fication Number	WT2	HT2	WT9	HT9	LG9	ST9	WT18	HT18	LG18	ST18	SOMA
201	13.6	90.2	41.5	139.4	31.6	74.0	110.2	179.0	44.1	226.0	7.0
202	12.7	91.4	31.0	144.3	26.0	73.0	79.4	195.1	36.1	252.0	4.0
203	12.6	86.4	30.1	136.5	26.6	64.0	76.3	183.7	36.9	216.0	6.0
204	14.8	87.6	34.1	135.4	28.2	75.0	74.5	178.7	37.3	220.0	3.0
205	12.7	86.7	24.5	128.9	24.2	63.0	55.7	171.5	31.0	200.0	1.5
206	11.9	88.1	29.8	136.0	26.7	77.0	68.2	181.8	37.0	215.0	3.0
207	11.5	82.2	26.0	128.5	26.5	45.0	78.2	172.5	39.1	152.0	6.0
209	13.2	83.8	30.1	133.2	27.6	70.0	66.5	174.6	37.3	189.0	4.0
210	16.9	91.0	37.9	145.6	29.0	61.0	70.5	190.4	33.9	183.0	3.0
211	12.7	87.4	27.0	132.4	26.0	74.0	57.3	173.8	33.3	193.0	3.0
212	11.4	84.2	25.9	133.7	25.8	68.0	50.3	172.6	31.6	202.0	3.0
213	14.2	88.4	31.1	138.3	27.3	59.0	70.8	185.2	36.6	208.0	4.0
214	17.2	87.7	34.6	134.6	30.6	87.0	73.7	178.4	39.2	227.0	3.0
215	13.7	89.6	34.6	139.0	28.9	71.0	75.2	177.6	36.8	204.0	2.5
216	14.2	91.4	43.1	146.0	32.4	98.0	83.1	183.5	38.0	226.0	4.0
217	15.9	90.0	33.2	133.2	28.5	82.0	74.3	178.1	37.8	233.0	2.5
218	14.3	86.4	30.7	133.3	27.3	73.0	72.2	177.0	36.5	237.0	2.0
219	13.3	90.0	31.6	130.3	27.5	68.0	88.6	172.9	40.4	230.0	7.0
221	13.8	91.4	33.4	144.5	27.0	92.0	75.9	188.4	36.5	250.0	1.0
222	11.3	81.3	29.4	125.4	27.7	70.0	64.9	169.4	35.7	236.0	3.0
223	14.3	90.6	30.2	135.8	26.7	70.0	65.6	180.2	35.4	177.0	4.0
224	13.4	92.2	31.1	139.9	27.2	63.0	66.4	189.0	35.3	186.0	4.0
225	12.2	87.1	27.6	136.8	25.8	73.0	59.0	182.4	33.5	199.0	3.0
226	15.9	91.4	32.3	140.6	27.9	69.0	68.1	185.8	34.2	227.0	1.0
227	11.5	89.7	29.0	138.6	24.6	61.0	67.7	180.7	34.3	164.0	4.0
228	14.2	92.2	31.4	140.0	28.2	74.0	68.5	178.7	37.0	219.0	2.0

X_8 = HT18 = height at age 18.

X_9 = LG18 = leg circumference.

X_{10} = ST18 = strength at age 18.

X_{11} = SOMA = somatotype, a seven-point scale, as a measure of fatness (1 = slender, 7 = fat), determined using a photograph taken at age 18.

Data for 26 boys and for 32 girls are given in Tables 2.6 and 2.7, respectively (the complete study consisted of larger sample sizes and of more variables; see Tuddenham and Snyder (1954) for details).

Table 2.7   Berkeley Guidance Study: girls

Identi-fication Number	WT2	HT2	WT9	HT9	LG9	ST9	WT18	HT18	LG18	ST18	SOMA
331	12.6	83.8	33.0	136.5	29.0	57.0	71.2	169.6	38.8	107.0	6.0
334	12.0	86.2	34.2	137.0	27.3	44.0	58.2	166.8	34.3	130.0	5.0
335	10.9	85.1	28.1	129.0	27.4	48.0	56.0	157.1	37.8	101.0	5.0
351	12.7	88.6	27.5	139.4	25.7	68.0	64.5	181.1	34.2	149.0	4.0
352	11.3	83.0	23.9	125.6	24.5	22.0	53.0	158.4	32.4	112.0	5.0
353	11.8	88.9	32.2	137.1	28.2	59.0	52.4	165.6	33.8	136.0	4.0
354	15.4	89.7	29.4	133.6	26.6	58.0	56.8	166.7	32.7	118.0	4.5
355	10.9	81.3	22.0	121.4	24.4	44.0	49.2	156.5	33.5	110.0	4.0
356	13.2	88.7	28.8	133.6	26.5	58.0	55.6	168.1	34.1	104.0	4.5
357	14.3	88.4	38.8	134.1	31.1	57.0	77.8	165.3	39.8	138.0	6.5
358	11.1	85.1	36.0	139.4	28.2	64.0	69.6	163.7	38.6	108.0	5.5
359	13.6	91.4	31.3	138.1	27.6	64.0	56.2	173.7	34.2	134.0	3.5
361	13.5	86.1	33.3	138.4	29.4	73.0	64.9	169.2	36.7	141.0	5.0
362	16.3	94.0	36.2	139.5	28.0	52.0	59.3	170.1	32.8	122.0	4.5
364	10.2	82.2	23.4	129.8	22.6	60.0	49.8	164.2	30.0	128.0	4.0
365	12.6	88.2	33.8	144.8	28.3	107.0	62.6	176.0	35.8	168.0	5.0
366	12.9	87.5	34.5	138.9	30.5	62.0	66.6	170.9	38.8	126.0	5.0
367	13.3	88.6	34.4	140.3	31.2	88.0	65.3	169.2	39.0	142.0	5.0
368	13.4	86.9	38.2	143.8	29.8	78.0	65.9	172.0	35.7	132.0	5.5
369	12.7	86.4	31.7	133.6	27.5	52.0	59.0	163.0	32.7	116.0	5.5
370	12.2	80.9	26.6	123.5	27.2	40.0	47.4	154.5	32.2	112.0	4.0
371	15.4	90.0	34.2	139.9	29.1	71.0	60.4	172.5	35.7	137.0	4.0
372	12.7	94.0	27.7	136.1	26.7	30.0	56.3	175.6	34.0	114.0	3.0
373	13.2	89.7	28.5	135.8	25.5	76.0	61.7	167.2	35.5	122.0	4.5
374	12.4	86.4	30.5	131.9	28.6	59.0	52.4	164.0	34.8	121.0	5.0
376	13.4	86.4	39.0	130.9	29.3	38.0	58.4	161.6	33.0	107.0	6.5
377	10.6	81.8	25.0	126.3	25.0	50.0	52.8	153.6	33.4	140.0	5.0
380	12.7	91.4	29.8	135.5	27.0	57.0	67.4	173.5	34.5	123.0	5.0
382	11.8	88.6	27.0	134.0	26.5	54.0	56.3	166.2	36.2	135.0	4.5
383	13.3	86.4	41.4	138.2	32.5	44.0	82.8	162.8	42.5	125.0	7.0
384	13.2	94.0	41.6	142.0	31.0	56.0	68.1	168.6	38.4	142.0	5.5
385	15.9	89.2	42.4	140.8	32.6	74.0	63.1	169.2	37.9	142.0	5.5

2.1.1. For the girls, obtain the matrix of sample correlations, the sample means, and the sample standard deviations.

2.1.2. Fit the model

$$\text{SOMA} = \beta_0 + \beta_1\,\text{HT2} + \beta_2\,\text{WT2} + \beta_3\,\text{HT9} + \beta_4\,\text{WT9} + \beta_5\,\text{ST9} + e$$

Find $\hat{\sigma}$, R^2, the overall analysis of variance and F-test, and state the conclusion for the F-test. Compute t-statistics to be used to test each of the β_j in the model equal to zero. Explicitly state the hypothesis tested in each, and the conclusion.

2.1.3. Now fit the model

$$\text{SOMA} = \beta_0 + \beta_3\,\text{HT9} + \beta_4\,\text{WT9} + \beta_5\,\text{ST9} + e$$

and compare this model to that fit in 2.1.2 (that is, compute an F-test).

2.1.4. Repeat 2.1.1–2.1.3, except for the boys. Qualitatively describe the difference between the fitted models for boys and for girls (formal procedures will be studied in Chapter 7).

2.1.5. For the boys, repeat the derivation in Section 2.1 when adding the variable WT9 to the model $\text{HT18} = \beta_0 + \beta_1\,\text{WT2}$.

2.2   (Matrix manipulation). Define the following matrices:

$$\mathbf{A} = \begin{bmatrix} 1 & 1 \\ -1 & 0 \end{bmatrix} \quad \mathbf{B} = \begin{bmatrix} 3 & 1 \\ 2 & 1 \end{bmatrix} \quad \mathbf{I} = \begin{bmatrix} 1 & 0 \\ 0 & 1 \end{bmatrix} \quad \mathbf{D} = \begin{bmatrix} -2 \\ 3 \end{bmatrix}$$

$$\mathbf{E} = \begin{bmatrix} 2 \\ 1 \end{bmatrix} \quad \mathbf{H} = \begin{bmatrix} \dfrac{1}{\sqrt{2}} & \dfrac{-1}{\sqrt{2}} \\ \dfrac{1}{\sqrt{2}} & \dfrac{1}{\sqrt{2}} \end{bmatrix} \quad \mathbf{C} = \begin{bmatrix} 1 & 2 \\ 3 & 4 \\ 5 & 6 \end{bmatrix}$$

2.2.1. Find $\mathbf{A}^T$, $\mathbf{B}^T$, $\mathbf{C}^T$, $\mathbf{D}^T$, $\mathbf{E}^T$.

2.2.2. Find $\mathbf{A} + \mathbf{B}$.

2.2.3. Find $\mathbf{AB}$ and $\mathbf{BA}$. Does $\mathbf{AB} = \mathbf{BA}$?

2.2.4. Show that $(\mathbf{AB})^T = \mathbf{B}^T\mathbf{A}^T$.

2.2.5. Compute $\mathbf{C}^T\mathbf{C}$ and $\mathbf{CC}^T$. Are they equal?

2.2.6. Find $\mathbf{DE}^T$, $\mathbf{D}^T\mathbf{E}$.

2.2.7. Show that $\mathbf{H}$ is orthogonal (that is, $\mathbf{HH}^T = \mathbf{H}^T\mathbf{H} = \mathbf{I}$).

2.3   Partitioned matrices. An $n \times p$ matrix $\mathbf{C}$ may be partitioned by columns into $\mathbf{C} = (\mathbf{C}_1\ \mathbf{C}_2)$, where $\mathbf{C}_1$ is an $n \times p_1$ matrix and $\mathbf{C}_2$ is an $n \times (p - p_1)$ matrix, and $\mathbf{C}_1$ is the first p_1 columns of $\mathbf{C}$, and $\mathbf{C}_2$ is the last

$p - p_1$ columns of $\mathbf{C}$. With this definition

$$\mathbf{C}^T\mathbf{C} = (\mathbf{C}_1 \ \mathbf{C}_2)^T(\mathbf{C}_1 \ \mathbf{C}_2)$$

$$= \begin{pmatrix} \mathbf{C}_1^T \\ \mathbf{C}_2^T \end{pmatrix}(\mathbf{C}_1 \ \mathbf{C}_2)$$

$$= \begin{pmatrix} \mathbf{C}_1'\mathbf{C}_1 & \mathbf{C}_1^T\mathbf{C}_2 \\ \mathbf{C}_2^T\mathbf{C}_1 & \mathbf{C}_2^T\mathbf{C}_2 \end{pmatrix}$$

and

$$\mathbf{C}\mathbf{C}^T = (\mathbf{C}_1 \ \mathbf{C}_2)(\mathbf{C}_1 \ \mathbf{C}_2)^T = \mathbf{C}_1\mathbf{C}_1^T + \mathbf{C}_2\mathbf{C}_2^T$$

2.3.1. Show, by direct multiplication, that if $(\mathbf{C}^T\mathbf{C})$ is of full rank, we can write

$$(\mathbf{C}^T\mathbf{C})^{-1} = \begin{bmatrix} (\mathbf{C}_1^T\mathbf{C}_1)^{-1} + \mathbf{FE}^{-1}\mathbf{F}^T & -\mathbf{FE}^{-1} \\ -\mathbf{E}^{-1}\mathbf{F}^T & \mathbf{E}^{-1} \end{bmatrix}$$

where

$$\mathbf{E} = \mathbf{C}_2^T\mathbf{C}_2 - \mathbf{C}_2^T\mathbf{C}_1(\mathbf{C}_1^T\mathbf{C}_1)^{-1}\mathbf{C}_1^T\mathbf{C}_2,$$

and

$$\mathbf{F} = (\mathbf{C}_1^T\mathbf{C}_1)^{-1}\mathbf{C}_1^T\mathbf{C}_2$$

2.3.2. Suppose that the correct linear model is $\mathbf{Y} = \mathbf{X}\boldsymbol{\beta} + \mathbf{e}$, with $\mathbf{X}$ an $n \times p'$ matrix, but we fit the model $\mathbf{Y} = \mathbf{X}\boldsymbol{\beta} + \mathbf{Z}\boldsymbol{\gamma} + \mathbf{e}$, so that, unknown to us, $\boldsymbol{\gamma} = \mathbf{0}$. Use the result of 2.3.1 to show that the estimate of $\boldsymbol{\beta}$ in the latter model is unbiased, and find its variance. Compare the variance of $\hat{\boldsymbol{\beta}}$ from the correct linear model to the variance that would be obtained if the larger model were fit, and comment. Find conditions on $\mathbf{X}$ and $\mathbf{Z}$ such that the estimator for $\boldsymbol{\beta}$ is numerically identical for either model.

2.4   QR decomposition. Suppose we have an $n \times p'$ matrix $\mathbf{Q}_1$ and a $p' \times p'$ upper triangular matrix $\mathbf{R}$ such that $\mathbf{Q}_1^T\mathbf{Q}_1 = \mathbf{I}_{p'}$, and $\mathbf{Q}_1\mathbf{R} = \mathbf{X}$ (as in Appendix 2A.3).

2.4.1. Show that $\mathbf{R}^T\mathbf{R} = \mathbf{X}^T\mathbf{X}$.
2.4.2. Verify (2A.16).
2.4.3. Show that $(\mathbf{X}^T\mathbf{X})^{-1}\mathbf{X}^T\mathbf{Y} = \mathbf{R}^{-1}\mathbf{Q}_1^T\mathbf{Y}$.
2.4.4. Using $\mathbf{Q}_2$ as defined in Section 2A.3, it is convenient to define $\mathbf{z}_1 = \mathbf{Q}_1^T\mathbf{Y}$ and $\mathbf{z}_2 = \mathbf{Q}_2^T\mathbf{Y}$. Then from (2A.16), $RSS = \mathbf{z}_2^T\mathbf{z}_2$. Show that RSS

can also be written as

$$RSS = \mathbf{Y}^T\mathbf{Y} - \mathbf{z}_1^T\mathbf{z}_1$$

This latter form is often convenient for computing.

2.4.5. Show that the vector of fitted values $\hat{\mathbf{Y}} = \mathbf{Q}_1\mathbf{z}_1$. This, incidentally, will show that $\mathbf{Q}_1\mathbf{Q}_1^T = \mathbf{X}(\mathbf{X}^T\mathbf{X})^{-1}\mathbf{X}^T$, which will be an important matrix in Chapters 5 and 6. Also, find $\hat{\mathbf{e}}$ in terms of $\mathbf{Y}$, $\mathbf{Q}_1$, and $\mathbf{z}_1$.

2.4.6. We recall that the variance of a fitted value at any vector $\mathbf{x}$ is given by $\sigma^2\mathbf{x}^T(\mathbf{X}^T\mathbf{X})^{-1}\mathbf{x}$. Also, if we define $\mathbf{x}$ to be a unit vector (i.e., a vector with all zero entries except a 1 for the jth place), then $\sigma^2\mathbf{x}^T(\mathbf{X}^T\mathbf{X})^{-1}\mathbf{x}$ $= \text{var}(\hat{\beta}_j)$. Thus if we can find $\mathbf{x}^T(\mathbf{X}^T\mathbf{X})^{-1}\mathbf{x}$ for any $p' \times 1$ vector $\mathbf{x}$, we can find the estimated variance of any fitted value, prediction or coefficient.

From 2.4.1 above, since $\mathbf{X}^T\mathbf{X} = \mathbf{R}^T\mathbf{R}$, then $(\mathbf{X}^T\mathbf{X})^{-1} = (\mathbf{R}^T\mathbf{R})^{-1} = \mathbf{R}^{-1}\mathbf{R}^{-T}$, where $-T$ means inverse of the transpose.

Show that

$$\mathbf{x}^T(\mathbf{X}^T\mathbf{X})^{-1}\mathbf{x} = (\mathbf{R}^{-T}\mathbf{x})^T(\mathbf{R}^{-T}\mathbf{x})$$

Suppose we let $\mathbf{c} = \mathbf{R}^{-T}\mathbf{x}$. Then $\mathbf{x}^T(\mathbf{X}^T\mathbf{X})^{-1}\mathbf{x} = \mathbf{c}^T\mathbf{c}$. To find $\mathbf{c}$, write

$$\mathbf{c} = \mathbf{R}^{-T}\mathbf{x}$$

or

$$\mathbf{R}^T\mathbf{c} = \mathbf{x}$$

and use back substitution (Appendix 2A.3) to solve for $\mathbf{c}$. This avoids explicit inversion of $\mathbf{X}^T\mathbf{X}$ or of $\mathbf{R}$.

2.4.7. Suppose $p' = 3$ and

$$\mathbf{R} = \begin{bmatrix} 2 & 4 & 3 \\ 0 & 1 & 5 \\ 0 & 0 & 8 \end{bmatrix}$$

Assuming $\hat{\sigma}^2 = 1$, find $\text{var}(\hat{\beta}_0)$ (e.g., set $\mathbf{x} = (1, 0, 0)^T$), $\text{var}(\hat{\beta}_1)$ (set $\mathbf{x} = (0, 1, 0)^T$) and the covariance between $\hat{\beta}_0$ and $\hat{\beta}_1$. The latter will require a slight extension of the above result.

2.5   Computational example. Consider the linear model $\mathbf{Y} = \mathbf{X}\boldsymbol{\beta} + \mathbf{e}$, where

$$\mathbf{X} = \begin{bmatrix} 1 & 1 \\ 1 & 3 \\ 1 & 5 \\ 1 & 7 \end{bmatrix} \qquad \mathbf{Y} = \begin{bmatrix} 34 \\ 47 \\ 55 \\ 64 \end{bmatrix}$$

2.5.1. Compute $\mathbf{X}^T\mathbf{X}$, $\mathbf{X}^T\mathbf{Y}$, $\mathbf{Y}^T\mathbf{Y}$. Using the rule that if $\mathbf{A}$ is a 2×2

symmetric matrix, then, if $ac \neq b^2$,

$$\mathbf{A}^{-1} = \begin{pmatrix} a & b \\ b & c \end{pmatrix}^{-1} = \frac{1}{ac - b^2} \begin{pmatrix} c & -b \\ -b & a \end{pmatrix}$$

find $(\mathbf{X}^T\mathbf{X})^{-1}$, $\hat{\beta}$, and $\mathrm{var}(\hat{\beta})$. Find $\hat{\mathbf{Y}}$ and $\hat{\mathbf{e}}$.

2.5.2. Define

$$\mathbf{Q}_1 = \begin{bmatrix} -\dfrac{1}{2} & \dfrac{3\sqrt{5}}{10} \\[2ex] -\dfrac{1}{2} & \dfrac{\sqrt{5}}{10} \\[2ex] -\dfrac{1}{2} & -\dfrac{\sqrt{5}}{10} \\[2ex] -\dfrac{1}{2} & -\dfrac{3\sqrt{5}}{10} \end{bmatrix} \qquad \mathbf{R} = \begin{pmatrix} -2 & -10 \\ 0 & 2\sqrt{5} \end{pmatrix}$$

Show that $\mathbf{Q}_1^T\mathbf{Q}_1 = \mathbf{I}_2$ and $\mathbf{Q}_1\mathbf{R} = \mathbf{X}$. Show that $\mathbf{R}^T\mathbf{R} = \mathbf{X}^T\mathbf{X}$. Find $\mathbf{z}_1 = \mathbf{Q}_1^T\mathbf{Y}$ and $\hat{\mathbf{Y}} = \mathbf{Q}_1\mathbf{z}_1$. Compute $\hat{\mathbf{e}}$. Find $\hat{\beta}$ via back substitution.

3

DRAWING CONCLUSIONS

After the calculations are completed, the results must be interpreted. Although the outline of the analysis may be similar in many problems, the conclusions that may be drawn are varied. Often, a fitted model is an approximation, either because some variables are unmeasured or incorrectly measured, or because the functional form used is not exactly correct.

3.1 Interpreting parameter estimates

Do parameters exist? In many regression problems, the model $Y = X\beta + e$ is a useful fiction, suggested so that straightforward techniques for data analysis can be applied. As a result, $\hat{\beta}$ may not estimate a real quantity and, if data were collected on the same variables, but over different ranges, the computed $\hat{\beta}$ could "estimate" something else. Estimation of parameters, unknown constants that characterize a process, makes sense only when the presumed functional form of the model is nearly exact. In general, the computed value for $\hat{\beta}$ depends both on the underlying process that ties the variables together, and on the details of the data collection. Depending on these details, interpretation of parameter estimates is often difficult.

58

Interpreting estimates: magnitude. In the fuel consumption data of Example 2.1, the fitted model was

$$\widehat{\text{FUEL}} = 377.3 - 34.79 \text{ TAX} + 1336 \text{ DLIC}$$
$$- 0.0666 \text{ INC} - 0.00243 \text{ ROAD} \qquad (3.1)$$

The usual interpretation of a parameter estimate is as a rate of change: increasing TAX rate by 1¢ should decrease consumption, all other factors being held constant, by 34.79 gallons per person. Fundamental to this interpretation are the assumptions that a variable can in fact be changed by one unit without affecting the other variables, and that the model fit to the available data will apply when one of the variables is thus changed. In this example, the data are observational (i.e., assignment of values for the independent variables, such as TAX rate, was not under the control of the analyst), so whether or not fuel consumption would be decreased by increasing taxes cannot be directly assessed. Rather, a more conservative interpretation is in order: states with higher tax rates are observed to have lower fuel consumption. To draw conclusions concerning the effects of changing tax rates, the rates must in fact be changed, and the results observed.

Signs of estimates. The sign of the parameter estimates indicate the direction of the relationship between the predictor and the response. In multiple regression, if the independent variables are correlated, the signs for some parameter estimates may be opposite to an investigator's understanding of a problem. Alternatively, the sign of a fitted coefficient may differ depending on the variables included in a model. Thus in simple regression an estimated slope may be positive (the sample correlation between the two variables ignoring all others is positive) while in a multiple regression, the corresponding estimated slope may be negative (the sample partial correlation between the two variables adjusted for the others is negative). While this is mathematically reasonable, and occasionally scientifically reasonable, it is difficult to interpret the role of such a predictor. If the independent variables are changed by a linear transformation, more reasonable results can sometimes be obtained.

Example 3.1 Berkeley Guidance Study

Data from the Berkeley Guidance Study on the growth of boys and girls are given in Problem 2.1. In studying these data, suppose we wish to model somatotype (SOMA) by weights at ages 2, 9, and 18 (WT2, WT9, WT18) for $n = 32$ girls. The correlation matrix for these four variables is given in Table 3.1. All the variables are positively correlated, as one might expect. Yet, the regression of SOMA on WT2, WT9, and WT18, Table 3.2, leads to the unexpected conclusion that heavier girls at age 2 tend to be thinner (have lower somatotype) at age 18. This result may be due to the correlations between the independent variables. In place of these above variables, consider the following:

WT2 = Weight at age 2.

DW9 = WT9 − WT2 = weight gain from age 2 to 9.

DW18 = WT18 − WT9 = weight gain from age 9 to 18.

Since all three variables measure weights, combining them in this

Table 3.1   Correlation matrix for weight variables for Berkeley Guidance Study girls

WT2	1.00			
WT9	0.5969	1.000		
WT18	0.3508	0.7108	1.000	
SOMA	0.1234	0.6508	0.6865	1.000
	WT2	WT9	WT18	SOMA

Table 3.2   Regression of SOMA on WT2, WT9, and WT18

Variable	Coefficient	Standard Error	t-Value
Intercept	2.03686	1.08219	1.88
WT2	− 0.217635	$0.881797E-01$	− 2.47
WT9	$0.945833E-01$	$0.318707E-01$	2.97
WT18	$0.432694E-01$	$0.183939E-01$	2.35

$\hat{\sigma}^2 = 0.332625$, d.f. = 28, $R^2 = 0.610$

Table 3.3   Regression of SOMA on WT2, DW9, and DW18

Variable	Coefficient	Standard Error	t-Value
Intercept	2.03686	1.08219	1.88
WT2	− $0.797820E-01$	$0.761845E-01$	− 1.05
DW9	0.137853	$0.239081E-01$	5.77
DW18	$0.432694E-01$	$0.183939E-01$	2.35

$\hat{\sigma}^2 = 0.332625$, d.f. = 28, $R^2 = 0.610$

way is reasonable. However, if the variables measured different quantities, then combining them could lead to conclusions that are even less useful than those originally obtained. The fitted regression for SOMA on WT2, DW9, and DW18 is given in Table 3.3.

Compare this regression to that fit to the weights themselves. The estimates $\hat{\beta}_0$, se($\hat{\beta}_0$), $\hat{\sigma}^2$, and R^2 are identical in each. In fact, since the three variables WT2, DW9, and DW18 can be obtained from WT2, WT9, and WT18 via a (nonsingular) linear transformation, the two sets of variables carry exactly the same information concerning SOMA. However, the estimated coefficient for WT2 depends on which set of variables is used. Table 3.2 shows that $\hat{\beta}_{WT2} = -0.22$, with $t = -2.47$, while in Table 3.3, $\hat{\beta}_{WT2} = -0.08$, with $t = -1.05$. In the former case, the effect of WT2 appears substantial, while in the latter it does not. Although $\hat{\beta}_{WT2}$ is negative in each, we would be led in the latter case to conclude that the effect of WT2 is negligible. Thus interpretation of the effect of a variable depends not only on the other variables in a model, but also upon which linear transformation of those variables is used.

The linear transformation used above is not unique, and, depending on the context, many others might be preferred. For example, another set might be

$$AVE = (WT2 + WT9 + WT18)/3.$$

$$LIN = WT2 - WT18.$$

$$QUAD = WT2 - 2WT9 + WT18.$$

This transformation focuses on the fact that WT2, WT9, and WT18 are ordered in time and are more or less equally spaced. Pretending that the weight measurements are equally spaced, AVE, LIN, and QUAD are, respectively, the average, linear, and quadratic time trends in weight gain.

Tests. Even if the fitted model were correct and errors were normally distributed, tests and confidence statements for parameters are difficult to interpret because nonorthogonality in the data leads to a multiplicity of possible tests. Sometimes, tests of effects adjusted for other variables are clearly desirable, such as in assessing a treatment effect after adjusting for other variables to reduce variability. At other times, the order of fitting is not clear, and the analyst must expect ambiguous results.

If the model that is fit depends on the data, the situation is further complicated, since testing a fictional parameter equal to zero is not necessarily equivalent to testing the significance of a variable in a regression. However, the usual test statistics do provide a useful guide, even if strict probability interpretations are not reliable. For example, the t-value for WT2 in Table 3.3 indicates that WT2 has little effect on the regression, and essentially the same results would be obtained with or without that variable in the model.

All that has been said so far should serve to point out the futility of using accept/reject rules to determine significance of regressors. In most situations the only true test of significance is repeated experimentation.

3.2   Sampling models

Experimentation versus observation. There are fundamentally two types of independent variables in a regression analysis: experimental and observational. For the former type, the values of the independent variables are under the control of the experimenter, while for the latter type, the values of the independent variables are observed, rather than set. Consider, for example, a hypothetical study of factors determining the yield of a certain crop. Experimental variables might include the amount and type of fertilizers used, the spacing of plants, and the amount of irrigation, since each of these can be assigned by the investigator to the units, which are plots of land. Observational independent variables might include characteristics of the plots in the study, such as drainage, exposure, and soil fertility, and weather variables. All of these are beyond the control of the experimenter, yet may have important effects on the observed yields.

Some experimental designs, including those that use randomization, are constructed so that the effects of observational factors can be ignored or used in analysis of covariance (see, e.g., Cox (1958)). For data from designed experiments, fitted models may lead to useful results concerning the effects of experimental factors, and these models can be used in predicting future values of the response.

At the other extreme, purely observational studies not under the control of the analyst can only be used to predict or model the events that were observed in the data, as in the fuel consumption example. To apply observational results to predict future values, additional assumptions about the behavior of future values compared to the behavior of the existing data must be made.

Sampling from a normal population. Much of the intuition for the use of least squares estimation is based on the assumption that the observed data are a sample from a multivariate normal population. While the assumption of multivariate normality is almost never tenable in practical regression problems, it is worthwhile to explore the relevant results for normal data, first assuming random sampling, and then removing that assumption.

Example 3.2 Multivariate normality

Suppose that all of the observed variables, independent and dependent, are normal random variables, and the observations on each case are independent of the observations on each other case. In a two-variable problem, for the ith case observe (x_i, y_i), and suppose that

$$\begin{pmatrix} x_i \\ y_i \end{pmatrix} \sim N\left(\begin{pmatrix} \mu_X \\ \mu_Y \end{pmatrix}, \begin{pmatrix} \sigma_X^2 & \rho_{XY}\sigma_X\sigma_Y \\ \rho_{XY}\sigma_X\sigma_Y & \sigma_Y^2 \end{pmatrix} \right) \qquad i = 1, 2, \ldots, n \quad (3.2)$$

Equation (3.2) says that x_i and y_i are each realizations of normal random variables with means μ_X and μ_Y, variances σ_X^2 and σ_Y^2, and correlation coefficient ρ_{XY}. Now, suppose we consider the *conditional distribution* of y_i given that we have already observed the value of x_i. It can be shown (see, e.g., Lindgren (1976)) that the conditional distribution of y_i given x_i, written as $y_i \mid x_i$, is normal, and

$$y_i \mid x_i \sim N\left(\mu_y + \rho_{XY} \frac{\sigma_Y}{\sigma_X}(x_i - \mu_X), \sigma_Y^2(1 - \rho_{XY}^2) \right) \qquad i = 1, 2, \ldots, n$$

$$(3.3)$$

If we define

$$\beta_0 = \mu_Y - \beta_1 \mu_X; \qquad \beta_1 = \rho_{XY} \frac{\sigma_Y}{\sigma_X}; \qquad \sigma^2 = \sigma_Y^2(1 - \rho_{XY}^2) \quad (3.4)$$

then the conditional distribution of y_i given x_i is simply

$$y_i \mid x_i \sim N(\beta_0 + \beta_1 x_i, \sigma^2) \qquad i = 1, 2, \ldots, n \qquad (3.5)$$

which is essentially the same as the simple regression model.

Given random sampling, the five parameters in (3.2) are estimated, using the notation of Table 1.2, by

$$\hat{\mu}_X = \bar{x} \quad \hat{\sigma}_X^2 = s_X^2 \quad \hat{\rho}_{XY} = r_{XY}$$
$$\hat{\mu}_Y = \bar{y} \quad \hat{\sigma}_Y^2 = s_Y^2 \qquad (3.6)$$

Estimates of β_0 and β_1 are obtained by substituting estimates from (3.6) for parameters in (3.4), so that $\hat{\beta}_1 = r_{XY}s_Y/s_X$, and so on, as derived in Chapter 1. (However, $\hat{\sigma}^2 = [(n - 1)/(n - 2)]s_Y^2(1 - r_{XY}^2)$, to correct for degrees of freedom.)

For multiple regression, if the observations on the ith case are y_i and the $p' \times 1$ vector $\mathbf{x}_i$, and all are assumed to have normal distributions, and if $\sigma_Y^2 = \text{var}(y_i)$, $i = 1, 2, \ldots, n$, and $\mathcal{R}^2$ is the population multiple correlation between y_i and $\mathbf{x}_i$, then the conditional distribution of $y_i \,|\, \mathbf{x}_i$ is normal,

$$y_i \,|\, \mathbf{x}_i \sim N\left(\mathbf{x}_i^T \boldsymbol{\beta}, \sigma_Y^2(1 - \mathcal{R}^2)\right) \qquad i = 1, 2, \ldots, n \qquad (3.7)$$

where $\boldsymbol{\beta}$ is a function of the means, variances and covariances $\mathbf{x}_i$ and y_i.

Example 3.3   Nonrandom sampling of a population (or, how to get greater R^2).

The reader may have noticed that the conditional distribution in (3.3) or (3.7) does not depend on random sampling, but only on normal distributions. Thus whenever multivariate normality seems to be a reasonable model for the variables, a linear regression model is suggested for the conditional distribution of one variable given the others. However, if random sampling is not used, some of the usual summary statistics, including R^2, lose meaning. This is illustrated with artificial data.

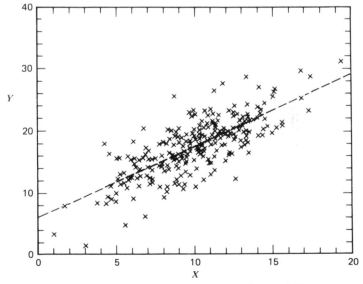

Figure 3.1   A bivariate normal sample, $n = 250$.

Figure 3.1 gives a bivariate pseudorandom sample of $n = 250$ pairs (x_i, y_i) drawn on a computer as if each (x_i, y_i) was independently drawn from

$$\binom{x_i}{y_i} \sim N\left(\binom{10.0}{17.5}, \binom{9.0000 \quad 11.2510}{11.2510 \quad 23.0651} \right) \qquad i = 1, 2, \ldots, 250$$

(3.8)

From (3.3), the conditional distribution of $y_i \mid x_i$ is

$$y_i \mid x_i \sim N(5. + 1.25x_i, 9) \qquad i = 1, 2, \ldots, 250$$

In the figure, the cloud of points is generally elliptical, as is characteristic of normal data. The fitted regression line, given in Table 3.4, and drawn in Figure 3.1, is not far from the actual line $y = 5. + 1.25x$. The computed $R^2 = 0.584$ is close to the true $\rho_{XY}^2 = 0.610$. All of the usual summaries and tests will be useful because random sampling was used.

Now, consider the same set of points, but now select cases according to their value x_i. Let $S_X^2 = \sum (x_i - \bar{x})^2 / n$ be the variance of the values of X that are actually in the sample; S_X^2 is not an estimate of σ_X^2 because the values of X were selected, not sampled. Often, the experimenter can choose units to make S_X^2 take on any value; usually, in designed experiments S_X^2 is as large as possible. Two selections are shown graphically in Figures 3.2 and 3.3. These two figures were obtained from Figure 3.1. In Figure 3.2, only cases with $x_i \leqslant 7$ or $x_i \geqslant 13$ were chosen, while in Figure 3.3, the range for X was restricted to $7 < x_i < 13$. Thus in Figure 3.2, S_X^2 is large, while in Figure 3.3, S_X^2 is small. For all three data sets (Figures 3.1–3.3) the fitted equations are nearly identical, and $\hat{\sigma}^2$ is fairly constant, as given in Table 3.4. However, note the very large change in R^2. In Figure 3.2, where $S_X^2 > \sigma_X^2$, $R^2 = 0.762$ is much too great, while in Figure 3.3, where $S_X^2 < \sigma_X^2$, $R^2 = 0.279$ is much too small. While all three graphs lead to nearly the same fitted equation, the usual summary of the fit—R^2—leads to very different conclusions.

Table 3.4 Regression summaries

			Estimates from	
	True Values	Full Sample	$X \leqslant 7$ or $X \geqslant 13$	$7 < X < 13$
n	∞	250	90	160
β_0	5.000	5.816	5.829	5.376
β_1	1.250	1.167	1.157	1.216
σ^2	9.000	9.010	9.166	9.024
R^2	0.610	0.584	0.762	0.279

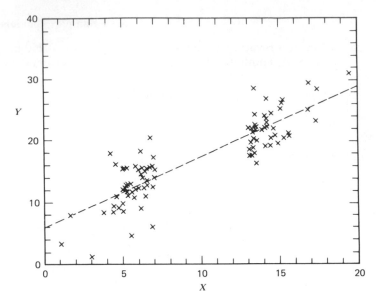

Figure 3.2 Sample with $x_i \leqslant 7$ or $x_i \geqslant 13$.

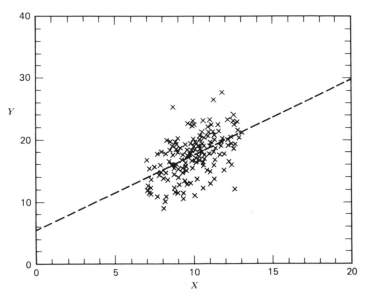

Figure 3.3 Sample with $7 \leqslant x_i \leqslant 13$.

The reason for this is apparent from the three graphs. In Figure 3.3, it appears that, because the range of X is so narrow, the variation in Y ignoring X (horizontal variation) is not much greater than the variation of Y given X (variability about the regression line). Thus the regression appears to explain little—and R^2 is small. A similar argument applied to Figure 3.2 suggests large R^2.

This example points out that even in the unusual event of analyzing data drawn from a multivariate normal population, if sampling of the population is not random, the interpretation of summary statistics such as R^2 may be completely misleading, as this statistic will be strongly influenced by the method of sampling. In particular, a few cases with unusual values for the predictors can largely determine the observed value of this statistic.

3.3   Independent variables measured with error

When the independent variables are measured with error, least squares analysis may be inappropriate since the desirable properties of least squares estimators, such as unbiasedness, may not hold. While alternative estimation methods are available, these generally require very strong assumptions or additional information concerning some of the parameters. Discussions of these methods are given by Acton (1959, Chapter 5) and Johnston (1963, Chapter 6).

The approach taken here is different. We shall use an index that will describe how much the parameter estimates would change if the independent variables were changed, or perturbed, by an amount thought to reflect the variability in the independent variables due to measurement error. If these changes are not too large, then it is reasonable to ignore measurement errors in the independent variables. However, if these changes are large, then alternatives to straightforward application of least squares must be used. This often requires a rethinking of the problem.

Notation. In the least squares problem, the estimated parameter vector is $\hat{\beta} = (\mathbf{X}^T\mathbf{X})^{-1}\mathbf{X}^T\mathbf{Y}$ with jth element $\hat{\beta}_j$, $j = 0, 1, 2, \ldots, p$, assuming a constant in the model. Next, suppose that $\mathbf{E}$ is an $n \times p'$ matrix of measurement errors with zero expectations, and define $\tilde{\mathbf{X}} = \mathbf{X} + \mathbf{E}$. It is convenient to view $\tilde{\mathbf{X}}$ as being obtained from $\mathbf{X}$ by perturbing it by adding $\mathbf{E}$, although one could also view $\tilde{\mathbf{X}}$ as the matrix of true values of the independent variables and $\mathbf{X} = \tilde{\mathbf{X}} - \mathbf{E}$ as the matrix of observed values with measurement error.

If we knew $\tilde{\mathbf{X}}$, we could compute $\tilde{\beta} = (\tilde{\mathbf{X}}^T\tilde{\mathbf{X}})^{-1}\tilde{\mathbf{X}}^T\mathbf{Y}$, the estimate of β

based on the perturbed data with jth element $\tilde{\beta}_j$, $j = 0, 1, \ldots, p$. Now suppose that σ_k is the standard deviation of the measurement error in the kth independent variable $k = 1, 2, \ldots, p$, which for now we assume is known, and define

$$
S = \begin{bmatrix} 0 & & & 0 \\ & \sigma_1^2 & & \\ & & \ddots & \\ 0 & & & \sigma_p^2 \end{bmatrix} \tag{3.9}
$$

to be the variance covariance matrix of the measurement errors in the independent variables, which are assumed to be uncorrelated. (If an intercept is not included in the model, then the first row and column should be removed from (3.9).) Hodges and Moore (1972) show that

$$
E(\hat{\beta}) - \beta = -(n - p')(X^T X)^{-1} S\beta \tag{3.10}
$$

so data with errors in the X's will lead to biased estimates. The bias depends on S, and as $n \to \infty$, it will approach a nonzero constant. Thus estimates based on perturbed data are also not consistent.

Monitoring the effects of perturbations. If $\tilde{X}$ were known, the effect of perturbing X to obtain $\tilde{X}$ could be measured by the ratios

$$
\frac{|\hat{\beta}_j - \tilde{\beta}_j|}{|\hat{\beta}_j|} \qquad j = 0, 1, 2, \ldots, p \tag{3.11}
$$

If the value of (3.11) is small, say about equal to 10^{-d}, then $\hat{\beta}_j$ and $\tilde{\beta}_j$ will agree to about d significant digits. For example, if it is equal to about $10^{-1} = 0.1$, $\hat{\beta}_j$ and $\tilde{\beta}_j$ will agree to about one significant digit, which will be an adequate agreement in many, but not all, regression problems. Of course, since $\tilde{X}$ is unknown, $\tilde{\beta}$ cannot be computed. Instead, upper bounds for (3.11) that depend on the σ_k's will be used.

Suppose that only the kth independent variable X_k is subject to measurement error, with standard deviation σ_k, and all other X's are measured without error. Then, following Stewart (1979), one can show that, for each $j = 0, 1, \ldots, p$, *

$$
|\hat{\beta}_j - \tilde{\beta}_j| \leqslant \gamma_{jk} \sigma_k \tag{3.12}
$$

where γ_{jk} is called a *sensitivity index* and is computed from

$$
\gamma_{jk} = \left(\hat{\beta}_k^2 c_{jj} + RSS c_{jk}^2 \right)^{1/2} \tag{3.13}
$$

---

* The bound will hold approximately if σ_k is relatively small and if the perturbations are independent. It is derived by expanding $\hat{\beta} - \tilde{\beta}$ in a series, and ignoring all terms after the first. One name for this method is *first order perturbation theory*.

where RSS is the residual sum of squares and c_{jj} and c_{jk} are the (j, j)-th and (j, k)-th elements of $(\mathbf{X}^T\mathbf{X})^{-1}$. Note the subscripts on γ_{jk}: the first subscript j refers to the element of $\boldsymbol{\beta}$ being studied $(j = 0, 1, \ldots, p)$ and the second subscript refers to the column of $\mathbf{X}$ subject to measurement error $(k = 1, 2, \ldots, p)$. Thus for each j we will have p upper bounds, depending on which column we assume is subject to measurement error. Joint effects of measurement errors in several variables are not considered in this development.

The relative sensitivity coefficient g_{jk} is defined

$$g_{jk} = \frac{\gamma_{jk}\sigma_k}{|\hat{\beta}_j|} \qquad \begin{array}{l} j = 0, 1, \ldots, p \\ k = 1, 2, \ldots, p \end{array} \qquad (3.14)$$

Dividing each side of (3.12) by $|\tilde{\beta}_j|$,

$$\frac{|\hat{\beta}_j - \tilde{\beta}_j|}{|\hat{\beta}_j|} \leqslant g_{jk} \qquad \begin{array}{l} j = 0, 1, \ldots, p \\ k = 1, 2, \ldots, p \end{array} \qquad (3.15)$$

Thus if all of the g_{jk}'s are sufficiently small, measurement error in the independent variables can be ignored.

Example 3.4 Fuel consumption data

To apply this method to the fuel consumption data, we need $(\mathbf{X}^T\mathbf{X})^{-1}$ from Table 2.3, $\hat{\boldsymbol{\beta}}$ from Table 2.4, and values for σ_{TAX}, σ_{DLIC}, σ_{INC}, and σ_{ROAD}. While exact values for these are unknown, we can guess at approximate values; generally, mistakes in estimating σ_k by a factor of 2 or 3 will not make much difference. For example, the choice of $\sigma_{\text{TAX}} = 0.1¢$, $\sigma_{\text{DLIC}} = 0.005$, $\sigma_{\text{INC}} = \$100$, and $\sigma_{\text{ROAD}} = 100$ miles may reflect possible measurement errors. $\sigma_{\text{TAX}} = 0.1¢$ was chosen to model the possibility that some states may have changed their tax rate during the year. Values for σ_{DLIC} and σ_{INC} were chosen to include uncertainties in measuring population and income. The value for σ_{ROAD} will probably depend on the miles of road in the state, but we make the simplifying assumption of applying a relatively large error to all states. These measurement errors are as large as one might expect in data like these, and, in practice, the true measurement errors may be smaller.

Computation of the g_{jk} of (3.14) is straightforward. For example, using results from Tables 2.3 and 2.4,

$$\gamma_{23} = \left[(-0.0666)^2(8.411) + 189050(-0.000145)^2\right]^{1/2} = 0.2032$$

and hence

$$g_{23} = \frac{\gamma_{23}\sigma_{INC}}{|\hat{\beta}_2|} \cong 1.5 \times 10^{-2}$$

Thus if X_3 = INC were replaced by another variable, based on X_3 but with random measurement error of magnitude σ_{INC} = \$100 added, the resulting $\tilde{\beta}_2$ for DLIC would agree with the computed $\hat{\beta}_2$ to about 2 or more significant digits. (This does *not* mean that $\hat{\beta}_2$ agrees with a "true" β_2 to 2 digits.) All the sensitivities are given in Table 3.5. Since they are all 10^{-1} or less and the assumed measurement error is relatively large, it is unlikely that measurement error in the X's will have much impact. However, if the assumed values for the σ_k's are correct, computed $\hat{\beta}_j$'s will have at most 2 or 3 reliable digits.

Table 3.5 The g_{jk}'s for the fuel consumption data

	Perturbed Column ($= k$)			
	1	2	3	4
Intercept	$5.6E-02$	$6.1E-02$	$5.2E-02$	$8.9E-03$
TAX	$5.2E-02$	$4.0E-02$	$3.8E-02$	$7.3E-03$
DLIC	$1.0E-02$	$2.0E-02$	$1.5E-02$	$1.4E-02$
INC	$1.4E-02$	$2.6E-02$	$5.1E-02$	$1.4E-03$
ROAD	$1.3E-01$	$1.5E-01$	$1.4E-01$	$4.7E-02$

Example 3.5 Longley data

The data given in Table 3.6 were first given by Longley (1967) to demonstrate inadequacies of various computer programs for regression calculations. The data consist of six independent variables and a response, as follows:

X_1 = GNP price deflator, in percent.

X_2 = Gross National Product (GNP), in millions of dollars.

X_3 = Unemployment, in thousands.

X_4 = Size of Armed Forces, in thousands.

X_5 = Noninstitutional population, 14 years of age and over, in thousands.

X_6 = Year.

Y = Total derived employment, in thousands.

Table 3.6 Longley's data

X_1	X_2	X_3	X_4	X_5	X_6	Y
83.0	234289.	2356.	1590.	107608.	1947.	60323.
88.5	259426.	2325.	1456.	108632.	1948.	61122.
88.2	258054.	3682.	1616.	109773.	1949.	60171.
89.5	284599.	3351.	1650.	110929.	1950.	61187.
96.2	328975.	2099.	3099.	112075.	1951.	63221.
98.1	346999.	1932.	3594.	113270.	1952.	63639.
99.0	365385.	1870.	3547.	115094.	1953.	64989.
100.0	363112.	3578.	3350.	116219.	1954.	63761.
101.2	397469.	2904.	3048.	117388.	1955.	66019.
104.6	419180.	2822.	2857.	118734.	1956.	67857.
108.4	442769.	2936.	2798.	120445.	1957.	68169.
110.8	444546.	4681.	2637.	121950.	1958.	66513.
112.6	482704.	3813.	2552.	123366.	1959.	68655.
114.2	502601.	3931.	2514.	125368.	1960.	69564.
115.7	518173.	4806.	2572.	127852.	1961.	69331.
116.9	554894.	4007.	2827.	130081.	1962.	70551.

Table 3.7 Regression for Longley's data

	Coefficient	Standard Error	t-Value
Intercept	$-0.348226E+07$	890420.	-3.91
X_1	15.0619	84.9149	0.18
X_2	$-0.358192E-01$	.334910$E-01$	-1.07
X_3	-2.02023	.488400	-4.14
X_4	-1.03323	.214274	-4.82
X_5	$-0.511041E-01$	.226073	$-.23$
X_6	1829.15	455.478	4.02

$\hat{\sigma}^2 = 92936.0$, d.f. $= 9$

Table 3.8 g_{jk} for Longley's data

	Perturbed Column ($= k$)					
	1	2	3	4	5	6
Intercept	$4.6E-03$	$7.5E-05$	$2.0E-03$	$9.2E-04$	$2.4E-04$	$5.7E-02$
$\hat{\beta}_1$	$4.7E-01$	$1.4E-03$	$4.0E-02$	$2.0E-02$	$8.3E-03$	$1.0E-00$
$\hat{\beta}_2$	$5.1E-02$	$3.3E-04$	$7.5E-03$	$3.3E-03$	$1.7E-03$	$2.0E-02$
$\hat{\beta}_3$	$1.1E-02$	$8.1E-05$	$2.0E-03$	$8.8E-04$	$4.1E-04$	$5.1E-02$
$\hat{\beta}_4$	$6.1E-03$	$4.0E-05$	$1.5E-03$	$8.3E-04$	$9.4E-05$	$4.0E-02$
$\hat{\beta}_5$	$2.4E-01$	$1.3E-03$	$3.3E-02$	$1.5E-02$	$9.9E-03$	$8.3E-01$
$\hat{\beta}_6$	$4.1E-03$	$7.2E-05$	$1.9E-03$	$8.9E-04$	$2.2E-04$	$5.6E-02$

The least squares computations for the model

$$Y = \beta_0 + \beta_1 X_1 + \beta_2 X_2 + \beta_3 X_3 + \beta_4 X_4 + \beta_5 X_5 + \beta_6 X_6 + e$$

are given in Table 3.7. Longley found that many computational algorithms could not produce this solution accurately, occasionally with no digits of agreement with the values in Table 3.7. In Table 3.8 the g_{jk}'s are given assuming $\sigma_1 = 0.1$, $\sigma_2 = \sigma_3 = \sigma_4 = \sigma_5 = 1$, and $\sigma_6 = 0.03$. These σ_k's represent round-off errors in the last digit, although the true measurement error is undoubtedly up to several thousand times larger. The measurement error for year, $\sigma_6 = 0.03$, represents variation of about 1 week. Actually, even this value is optimistically small, since the other variables were not measured exactly a year apart. Yet even with these tiny σ_k's, some of the g_{jk} (e.g., g_{16} and g_{56}) are near 10^{-1}, so *no more* than 1 significant digit in these β_j's is likely to be reliable. For this data set, any least squares calculations, accurate or inaccurate, are suspect, and meaningful interpretation of results is impossible.

The preceding might suggest that analysis is possible in the Longley data if the offending variable, in this case $X_6 =$ year, is deleted from the model. Since all computations are done relative to the whole model, this conclusion is not justified. More reasonable strategies for selection of variables are discussed in Chapter 8.

Problems

3.1. Fit the regression of SOMA on AVE LIN and QUAD as defined in Section 3.1 for the girls in the Berkeley Guidance Study data, and compare to the results in Section 3.1.

3.2. For a bivariate normal, find the conditional distribution of $x_i \mid y_i$, and note that the regression of Y on X is not the same, in general, as the regression of X on Y. Under what conditions is it the same? For the bivariate normal given at (3.8) find the conditional distribution of $x_i \mid y_i$, and draw the two regression lines (X on Y and Y on X) on the same graph.

3.3. For the Berkeley Guidance Study girls, compute the relative sensitivities for the model of Table 3.2. Explain your choice of σ_k's. What do you conclude about the sensitivity of the least squares analysis to errors in the X's?

4

WEIGHTED LEAST SQUARES, TESTING FOR LACK OF FIT, GENERAL *F*-TESTS, AND CONFIDENCE ELLIPSOIDS

This chapter begins with a discussion of the use of additional information about error variances and covariances. This information will sometimes be used to obtain generalized least squares estimators rather than the ordinary least squares estimators previously discussed. Alternatively, this information may be used to test for lack of fit of a model applied to observed data. Next, we turn to a heuristic discussion of the general methodology for obtaining test statistics that will follow F distributions when errors are normally distributed. These tests are used in many situations in regression, including those outlined in Chapters 1 and 2, as well as others to be described in later chapters. Finally, ellipsoids used as simultaneous confidence regions for more than one parameter are discussed. These regions, which depend on the F distribution, will be of use in assessing the influence of cases on regression estimates, a topic in Chapter 5.

4.1 Generalized and weighted least squares

The assumptions of earlier chapters that error variances are unknown and equal and that the errors are independent are usually made out of necessity since specific knowledge of these variances is exceptional. In a few problems, however, additional information concerning error variances will be available, with variances either known or known up to some multiplicative constant. The methodology for incorporating this information into the analysis is not difficult and is presented in this section.

Generalized least squares. Suppose that we know the value of a symmetric positive definite matrix Σ, such that the covariance matrix for the error vector e is given by $\text{var}(e) = \sigma^2 \Sigma$, with $\sigma^2 > 0$, but not necessarily known. We might reasonably expect that in these circumstances the ordinary least squares estimator of β, although still unbiased, will no longer be the minimum variance estimator, since it ignores obviously useful information. Formally, consider the model

$$Y = X\beta + e \qquad X: n \times p' \text{ rank } p' \tag{4.1a}$$

$$\text{var}(e) = \sigma^2 \Sigma \qquad \Sigma \text{ known, } \sigma^2 > 0 \text{ not necessarily known} \tag{4.1b}$$

We will continue to use the symbol $\hat{\beta}$ for the estimator of β, even though the estimate will be obtained via generalized, not ordinary, least squares. Once $\hat{\beta}$ is determined, the residuals $\hat{e}$ are given by $\hat{e} = Y - \hat{Y} = \hat{Y} - X\hat{\beta}$. The estimator $\hat{\beta}$ is chosen to minimize the generalized residual sum of squares,

$$RSS = e^T \Sigma^{-1} e \tag{4.2}$$

Roughly speaking, the use of the generalized residual sum of squares recognizes that some of the residuals, or fitting errors, are more important than are others. In particular, residuals corresponding to errors with a larger error variance will be less important in the computation of the generalized residual sum of squares. The generalized least squares estimator is given by

$$\hat{\beta} = (X^T \Sigma^{-1} X)^{-1} X^T \Sigma^{-1} Y \tag{4.3}$$

While this last equation can be found directly, it is convenient to transform the problem specified by model (4.1) to one that can be solved by ordinary least squares. Then, all of the results for ordinary least squares can be applied to generalized least squares problems.

Model (4.1) differs from the ordinary least squares model only in that $\text{var}(e) = \sigma^2 \Sigma$. Now, suppose that we could find an $n \times n$ matrix C, such that C is symmetric and $C^T C = CC^T = \Sigma^{-1}$ (and $C^{-1} C^{-T} = \Sigma$). Such a matrix C will be called the *square root* of Σ^{-1}. Then, using Appendix 2A.2

on random vectors, the variance-covariance matrix of $\mathbf{Ce}$ is given by

$$
\begin{aligned}
\mathrm{var}(\mathbf{Ce}) &= \mathbf{C}(\sigma^2\mathbf{\Sigma})\mathbf{C}^T \\
&= \sigma^2\mathbf{C}(\mathbf{C}^{-1}\mathbf{C}^{-T})\mathbf{C}^T \\
&= \sigma^2\mathbf{C}\mathbf{C}^{-1}\mathbf{C}^{-T}\mathbf{C}^T \\
&= \sigma^2\mathbf{I}_n
\end{aligned}
\tag{4.4}
$$

Multiplying both sides of equation (4.1a) by $\mathbf{C}$ gives

$$
\mathbf{CY} = \mathbf{CX}\beta + \mathbf{Ce}
\tag{4.5}
$$

Now, define $\mathbf{Z} = \mathbf{CY}$, $\mathbf{W} = \mathbf{CX}$, and $\mathbf{d} = \mathbf{Ce}$. Then equation (4.5) becomes

$$
\mathbf{Z} = \mathbf{W}\beta + \mathbf{d}
\tag{4.6}
$$

where, from (4.4), $\mathrm{var}(\mathbf{d}) = \sigma^2\mathbf{I}_n$, and in (4.6) β is exactly the same as β in (4.1). Model (4.6) can be solved using ordinary least squares. For example, the estimator of $\hat{\beta}$ in terms of $\mathbf{Z}$ and $\mathbf{W}$ is from (2.15).

$$
\hat{\beta} = (\mathbf{W}^T\mathbf{W})^{-1}\mathbf{W}^T\mathbf{Z}
$$

Substituting $\mathbf{X}$'s, $\mathbf{Y}$'s and $\mathbf{C}$'s for $\mathbf{W}$ and $\mathbf{Z}$, this last equation becomes

$$
\begin{aligned}
\hat{\beta} &= \left[(\mathbf{CX})^T(\mathbf{CX})\right]^{-1}(\mathbf{CX})^T(\mathbf{CY}) \\
&= (\mathbf{X}^T\mathbf{C}^T\mathbf{CX})^{-1}(\mathbf{X}^T\mathbf{C}^T\mathbf{CY}) \\
&= (\mathbf{X}^T\mathbf{\Sigma}^{-1}\mathbf{X})^{-1}(\mathbf{X}^T\mathbf{\Sigma}^{-1}\mathbf{Y})
\end{aligned}
$$

which is the estimator given at (4.3).

Thus a practical procedure for generalized least squares is to obtain $\mathbf{C}$, the square root of $\mathbf{\Sigma}^{-1}$, multiply the observed data vector $\mathbf{Y}$ and matrix $\mathbf{X}$ on the left by $\mathbf{C}$, and solve the resulting regression problem using ordinary least squares on the data so transformed. A numerical difficulty arises in actually finding $\mathbf{C}$. However, in the special case of weighted least squares, computing $\mathbf{C}$ from $\mathbf{\Sigma}$ is simple.

Weighted least squares. For weighted least squares, each element of $\mathbf{e}$ is uncorrelated with each other element, but the variances need not be all equal. Then, $\mathbf{\Sigma}$ is given by

$$
\mathbf{\Sigma} =
\begin{pmatrix}
w_1 & & & & \\
& w_2 & & & 0 \\
& & w_3 & & \\
& & & \ddots & \\
0 & & & & w_n
\end{pmatrix}
\tag{4.7}
$$

where σ^2 is generally an unknown parameter, but the $w_i > 0$ are known weights or relative precisions. Cases with large w_i are relatively more variable than are cases with small w_i. Since Σ is diagonal, so is Σ^{-1}, and

$$\Sigma^{-1} = \begin{pmatrix} \dfrac{1}{w_1} & & & \\ & \dfrac{1}{w_2} & & 0 \\ & & \ddots & \\ 0 & & & \dfrac{1}{w_n} \end{pmatrix} \tag{4.8}$$

From this last expression we see that

$$C = \begin{pmatrix} \dfrac{1}{\sqrt{w_1}} & & & \\ & \dfrac{1}{\sqrt{w_2}} & & 0 \\ & & \ddots & \\ 0 & & & \dfrac{1}{\sqrt{w_n}} \end{pmatrix} \tag{4.9}$$

and $C^T C = \Sigma^{-1}$. Then the matrix $W = CX$ and the vector $Z = CY$ are simply obtained, for regression with an intercept, as

$$W = \begin{pmatrix} \dfrac{1}{\sqrt{w_1}} & \dfrac{x_{11}}{\sqrt{w_1}} & \cdots & \dfrac{x_{1p}}{\sqrt{w_1}} \\ \dfrac{1}{\sqrt{w_2}} & \dfrac{x_{21}}{\sqrt{w_2}} & & \dfrac{x_{2p}}{\sqrt{w_2}} \\ \vdots & \vdots & & \vdots \\ \dfrac{1}{\sqrt{w_n}} & \dfrac{x_{n1}}{\sqrt{w_1}} & & \dfrac{x_{np}}{\sqrt{w_p}} \end{pmatrix} \qquad Z = \begin{pmatrix} \dfrac{y_1}{\sqrt{w_1}} \\ \\ \vdots \\ \\ \dfrac{y_n}{\sqrt{w_n}} \end{pmatrix} \tag{4.10}$$

An ordinary or unweighted least squares problem is obtained from a weighted least squares problem by dividing each of the observed values for the ith case, *including the column of 1's*, by the square root of the weight for that case. Ordinary least squares can then be applied to the transformed data to obtain estimates of parameters, $\hat{\sigma}^2$, tests, confidence intervals, and residuals in the usual way.

Applications of weighted least squares. To use weighted least squares, $\text{var}(e_i) = \sigma^2 w_i$ must be known, or at least the w_i must be known for each i. For example, suppose that the recorded values in a data set consist of means of several observations, and corresponding to x_i, we record y_i, the average of n_i observations all taken at x_i. Then, if all observations are independent with variance σ^2, $\text{var}(e_i) = \sigma^2/n_i$, and $w_i = 1/n_i$. Alternatively, $\text{var}(e_i)$ may be a function of some observable variable X_j, and perhaps $\text{var}(e_i) = \sigma^2 x_{ij}$. Then, $w_i = x_{ij}$. Or, different cases may be measured on different equipment of known but unequal precision. Finally, the values of $\text{var}(e_i)$ may be essentially known exactly by collecting very large samples or from a theoretical model, as illustrated in the following example.

Example 4.1 Strong interaction

The purpose of the experiment described here is to study the interactions of certain kinds of elementary particles in collision with proton targets (Weisberg *et al.*, 1978). The particles studied include the π^- meson and its antiparticle π^+. These are unstable and can be produced by high-energy accelerators. They are in a group of particles called hadrons, and have a positive or negative electrical charge equal in magnitude to the charge of the electron. Hadrons interact via the electromagnetic force that holds atoms together. In addition, and unlike the electron, they also interact via the so-called strong interaction force that holds nuclei together. Although the electromagnetic force is well understood, the strong interaction is still somewhat mysterious to physicists, and this experiment was designed to test certain theories of the nature of the strong interaction.

The scattering processes studied are denoted

$$a + p \rightarrow c + x$$

and may be represented by a diagram:

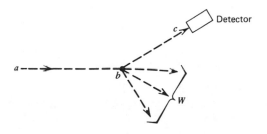

This diagram shows a beam of particles of type a aimed at a target b containing protons p. A detector then detects outgoing particles of type c. The experiment is carried out for a and c equal to various particles; only π^- and π^+ are considered here. For a typical collision at high energy, both a and p will break up and, by the process of the conversion of energy into matter, fragment into many hadrons. The experiment measures the rate of production of a particular particle type c, no matter what other particles (labeled W in the diagram) are produced.

The quantity measured is the scattering cross section y (usually denoted by $\Delta\sigma$), given by the equation

$$y = \frac{N_c}{N_a \rho l} \qquad (4.11)$$

where N_a is the number of incoming beam particles per second, ρ is the target density in particles per unit volume, l is the target length, and N_c is the number of particles c detected per second. The cross section y has the dimensions of area and is conventionally measured in millibarns (mb), where $1\text{ mb} = 1 \times 10^{-27}\text{ cm}^2$.

The experiment is carried out with beam a having various values of incident momentum p_a^{lab}, measured in the laboratory frame of reference. A quantity of more basic theoretical significance than p_a^{lab} is s, the square of the total energy in the center-of-mass frame of reference system. For the high momenta used in this experiment, particle a travels nearly at the speed of light, and the relationship

$$s = 2m_p p_a^{\text{lab}} \qquad (4.12)$$

holds to a good approximation. The units of s are $(\text{GeV})^2$, where 1 GeV $= 1 \times 10^9$ electron volts is the energy that an elementary particle reaches on being accelerated by an electric potential of one billion volts. The momentum p_a^{lab} and the mass m_p are measured in GeV, and $m_p = 0.938$ GeV for a proton.

Theoretical physicists have constructed various models of the strong interaction force. Certain models predict that, in the high-energy limit $s \to \infty$, the cross section y should approach a constant limit. In addition it is predicted that the functional form of the approach to this limit should be

$$y = \beta_0 + \beta_1 s^{-1/2} + \text{relatively small terms} \qquad (4.13)$$

The theory makes quantitative predictions about β_0 and β_1 and on their dependence on particle types a and c. Of interest, therefore, are: (1) estimation of β_0 and β_1, given (4.13) as a model, for each choice of a and c; (2) assessment of whether or not the model (4.13) provides an accurate description for the observed data; and (3) comparison of β_0 and β_1 to theoretical predictions if (4.13) is in fact appropriate.

The data given in Table 4.1 summarize the results of experiments with $a = c = \pi^-$. At each p_a^{lab}, a very large number of particles N_a was used so that the variance of the observed y values could be accurately obtained from theoretical considerations. The square roots of these variances (i.e., the $\sqrt{\sigma^2 w_i}$) are given in column 4 of Table 4.1.

Table 4.1   Data for the physics example

p_a^{lab} (GeV)	$s^{-1/2}$ (GeV)	y (μb)	$\sqrt{\sigma^2 w_i} =$ Estimated Standard Deviation
4	0.345	367	17
6	0.287	311	9
8	0.251	295	9
10	0.225	268	7
12	0.207	253	7
15	0.186	239	6
20	0.161	220	6
30	0.132	213	6
75	0.084	193	5
150	0.060	192	5

For continuity of notation, set $x = s^{-1/2}$, and $e =$ smaller terms, so (4.13) can be written as

$$y_i = \beta_0 + \beta_1 x_i + e_i \qquad i = 1, 2, \ldots, n \qquad (4.14)$$

with the e_i independent and $\text{var}(e_i) = \sigma^2 w_i$ as given in Table 4.1. Without any loss of generality, we can set $\sigma^2 = 1$, so the numbers in column 4 of Table 4.1 correspond to $\sqrt{w_i}$, $i = 1, 2, \ldots, n$.

The estimated values of β_0 and β_1 must be obtained by weighted least squares. This can be done either by directly minimizing the generalized residual sum of squares (4.2), which in scalar form is given by

$$\sum \frac{(y_i - \hat{\beta}_0 - \hat{\beta}_1 x_i)^2}{w_i} = \sum \frac{(y_i - \hat{y}_i)^2}{w_i} \qquad (4.15)$$

or by rescaling the data, and then applying ordinary least squares. In many computer programs the latter approach is used, and one specifies a column of numbers to be used as the weights. The program will then carry out the scaling, by dividing the data by the appropriate numbers. However, there is no consistency between programs, and some may multiply by the weights, or by their square roots. These alternatives are reasonable as they make cases with large case weights more important, rather than less important. If, for example, entries are to be multiplied by the case weights, then the problem considered here can be handled by defining a column of data to be the squared reciprocal of the numbers of column 4 of Table 4.1. Even computer programs that have no provision for weighted least squares may be used if transformations of data and regression through the origin are possible. First, divide each column of **X** and **Y** by the square root of the column of weights. Then, define a new variable that is equal to the reciprocal of the square root of the weights. The regression of the rescaled **Y** on the rescaled **X**'s and the new column just created, but through the origin, will give the correct least squares regression. The coefficient for the new column will be the correct intercept estimate.

Table 4.2 lists the results of the weighted least squares calculations, and the fitted line is plotted in Figure 4.1. The usual summaries such as R^2 and t-tests indicate that the fitted model matches the observed data reasonably well, and the parameter estimates themselves are well determined. The next question is whether or not (4.13) does in

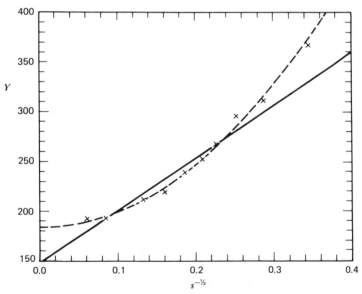

Figure 4.1   Scatter plot for the strong interaction data. Solid line: fitted simple regression; dashed line: fitted quadratic regression.

Table 4.2 Weighted least squares for the physics data

Variable	Estimate	Standard Error	t-Value
Intercept	148.4732	8.079	18.38
Slope	530.8354	47.550	11.16

$$\hat{\sigma}^2 = 2.744 \qquad R^2 = 0.94 \qquad \text{d.f.} = 8$$

Analysis of variance

Source	d.f.	SS	MS
Linear regression	1	342.0	342.0
Residual	8	21.95	2.744

fact fit the data. This question of fit or lack of fit of a model is the subject of the next section.

4.2 Testing for lack of fit, variance known

When the residual variance is known, or when an estimate of it can be obtained that does not depend on the presumed linear model, a test of lack of fit for the model can be made. A model specifies a shape for the relationship between the independent variables and the dependent variable. When the hypothesized shape is correct, then the residual mean square from the fitted model, $\hat{\sigma}^2$, will provide an unbiased estimate of the actual residual variance. However, if the hypothesized shape is incorrect, then $\hat{\sigma}^2$ will estimate a quantity larger than σ^2, since its size will depend both on the errors, and on systematic biases due to fitting the wrong shape. This suggests comparing the estimated $\hat{\sigma}^2$ from the model to the value of σ^2 known a priori or to an estimate that is valid regardless of the model. If $\hat{\sigma}^2$ is too large when compared to σ^2 or to its model-free estimate, we would have evidence that the fitted model does not account for some systematic features in the data, and the fitted model would not be adequate to describe the data.

In the strong interaction data (Example 4.1), we want to know if the straight-line model (4.13) or (4.14) provides an adequate description of the data. As outlined in the last section, the values in column 4 of Table 4.1 are used as weights, with $\sigma^2 = 1$, a known value. Also, from Table 4.2, $\hat{\sigma}^2 = 2.74$. Thus, we will reject the hypothesis that (4.13) or (4.14) provides an adequate model if we judge $\hat{\sigma}^2 = 2.74$ to be too large when compared to $\sigma^2 = 1$. To assign a p-value to this comparison we use the following result:

If the e_i's are independent, $N(0, \sigma^2 w_i)$, $i = 1, 2, \ldots, n$, with w_i known, and the linear model has parameters estimated using weighted least squares with the w_i's as weights, if the model is correct, then

$$\chi^2 = \frac{RSS}{\sigma^2} = \frac{(n - p')\hat{\sigma}^2}{\sigma^2} \tag{4.16}$$

is distributed as a chi-squared random variable with $n - p'$ degrees of freedom and, as usual, RSS is the residual sum of squares.

For the example, from Table 4.2,

$$\chi^2 = \frac{21.95}{1} = 21.95$$

From Table C at the back of the book $\chi^2(0.01, 8) = 20.09$, so the p-value associated with this test is less than 0.01, which suggests that this model may not be adequate for describing the data.

When this test indicates the failure of a model to give adequate fit, it is usual to fit alternative models. The common alternatives are either to transform some of the variables, as described in Chapter 6, or else to fit a model with higher order or polynomial terms in the independent variables. The available physical theory suggests this latter approach, and the model

$$y = \beta_0 + \beta_1 s^{-1/2} + \beta_2 (s^{-1/2})^2 + \text{smaller terms}$$

or

$$y_i = \beta_0 + \beta_1 x_i + \beta_2 x_i^2 + e_i \qquad i = 1, 2, \ldots, n \tag{4.17}$$

should be fit to the data. These models require that the relationship between y and x be a quadratic curve rather than a straight line.

To fit model (4.17), we are essentially fitting a multiple regression model with two regressors, x and x^2. Fitting must again use weighted least squares. The fitted equation and the analysis of variance are given in Table 4.3, and the fitted curve is graphed in Figure 4.1. The curve matches the data very closely. We can test for lack of fit of this model by computing

$$\chi^2 = \frac{RSS}{\sigma^2} = \frac{3.23}{1} = 3.23$$

Comparing this value to the percentage points of χ^2 with seven degrees of freedom clearly indicates that there is not evidence of lack of fit for this model.

Thus, although (4.13) does not adequately describe the data, (4.17) does result in an adequate fit. Judgement of the success or failure of the model for the strong interaction force requires analysis of data for other choices of incidence and product particles, as well as the data analyzed thus far. Based on this further analysis, Weisberg et al. (1978) concluded that the theoretical model for the strong interaction forces is consistent with the observed data.

Table 4.3   Model (4.17) for the physics data

Variable	Estimate	Standard Error	t-Value
Intercept	183.83	6.45	28.0
X	0.97	85.3	.01
X^2	1597.5	250.6	6.4

$\hat{\sigma}^2 = 0.46$ $R^2 = 0.99$ d.f. $= 7$

Analysis of variance

Source	d.f.	SS	MS
Regression	2	360.73	180.36
Residual	7	3.23	0.46

4.3   Testing for lack of fit, variance unknown

When the residual variance is unknown, a test for lack of fit of a model can be done if an estimate of σ^2 can be obtained that does not depend on whether or not the hypothesized model is correct. The most common way of finding this estimate is by repeated sampling of cases with identical independent variables. For example, consider the artificial data with $n = 10$ given in Table 4.4. These numbers were actually generated by first choosing the values of x_i and then computing $y_i = 2.0 + 0.5x_i + e_i$, $i = 1$, $2, \ldots, 10$, where the e_i were taken from a table of standard normal random deviates. If we consider only the values of y_i corresponding to $x_i = 1$, we can estimate the average response and standard deviation, with $3 - 1 = 2$ degrees of freedom, as shown. If we assume that the variance is constant for all values of x_i, a pooled estimate of this variance is obtained

Table 4.4   A hypothetical example

X	Y	$\bar{y}_i$	$\sum(y - \bar{y})^2$	s_i	d.f.
1	2.55 ⎫				
1	2.75 ⎬	2.6233	0.0243	0.1102	2
1	2.57 ⎭				
2	2.40 }	2.4000	0	0	0
3	4.19 ⎫	4.4450	0.1301	0.3606	1
3	4.70 ⎭				
4	3.81 ⎫				
4	4.87 ⎪	4.0325	2.2041	0.8571	3
4	2.93 ⎬				
4	4.52 ⎭				

SS(pe) $= 2.3585$,   d.f.(pe) $= 6$

by adding the sums of squares within group, often called the *sum of squares for pure error*, symbolically SS(pe), and dividing it by its degrees of freedom d.f.(pe), which is the sum of the degrees of freedom for each group. In Table 4.4,

$$SS(pe) = 0.0243 + 0.0000 + 0.1301 + 2.2041 = 2.3585$$

and the associated degrees of freedom is d.f.(pe) $= 2 + 0 + 1 + 3 = 6$, so the pooled estimate of variance, estimated without reference to the fitted model is $\hat{\sigma}^2_{pooled} = SS(pe)/d.f.(pe) = 0.3931$. It is important to note that if one were computing a one-way analysis of variance, with cases grouped according to their value of x_i, the mean square within groups would be identical to $\hat{\sigma}^2_{pooled}$.

Up to this point in the example the only assumption made is that of constant variance. Now, suppose we fit the linear regression model $y_i = \beta_0 + \beta_1 x_i + e_i$, $i = 1, 2, \ldots, 10$, as in Chapter 1. The analysis of variance table for this model is given in Table 4.5. The residual mean square in Table 4.5 provides an estimate of σ^2 that depends on the model. Thus we have two estimates of σ^2 and, if the latter is much larger than the former, the model does not fit.

Table 4.5   Analysis of variance for data of Table 4.4

Source	d.f.	SS	MS	F
Regression	1	4.2166	4.2166	
Residual	8	4.5693	0.5712	
{ Lack of fit	{ 2	{ 2.2108	{ 1.1054	2.81
{ Pure error	{ 6	{ 2.3585	{ 0.3931	

To obtain a test statistic that, under normality, is distributed as an F, the residual sum of squares in Table 4.5 is divided into the sum of squares for pure error, from Table 4.4, and the remainder, called the sum of squares for lack of fit or SS(lof), where SS(lof) $= RSS - SS(pe) = 4.5693 - 2.3585 = 2.2108$. The F-test for lack of fit of the model is then the ratio of the mean square for lack of fit to the mean square for pure error. The observed value of $F = 2.81$ is considerably smaller than $F(0.05; 2, 6) = 5.14$, suggesting no lack of fit of the model to these data.

Example 4.2   Apple shoots

Many types of trees produce two types of morphologically different shoots. Some branches remain vegetative year after year and contribute considerably to the size of the tree. Called long shoots or leaders, they may grow as much as 15 or 20 cm over a single growing season.

On the other hand, other shoots will seldom exceed 1 cm in total length. Called short, dwarf, or spur shoots, these usually produce flowers from which fruit may arise. To complicate the issue further, traditionally long shoots occasionally change to short in a new growing season and vice versa. The mechanism that the tree uses to control the long and short shoots is not well understood.

Bland (1978) has done a descriptive study of the difference between long and short shoots of McIntosh apple trees. Using healthy trees of clonal stock planted in 1933 and 1934, he took samples of long and short shoots from the trees every few days throughout the 1971 growing season (about 106 days). The shoots sampled are presumed to be a sample of available shoots at the sampling dates. The sampled shoots were removed from the tree, marked, and taken to the laboratory for analysis.

Among the many measurements taken, Bland counted the number of stem units in each shoot. The long and the short shoots could differ because of the number of stem units, the average size of stem units, or both. An abstract of Bland's data is given in Table 4.6, for both long and short shoots. For now we will consider only the long shoots, leaving the short shoots to the section of problems.

Our goal is to find an equation that can adequately describe the relationship between DAY = days from dormancy and Y = number of stem units. Lacking a theoretical form for this equation, we first examine Figure 4.2, a scatter plot of DAY versus average number of

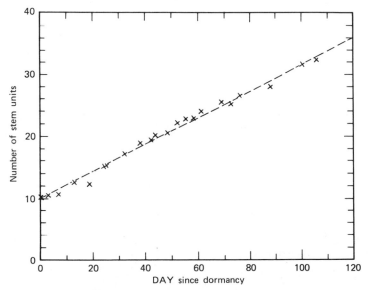

Figure 4.2 Scatter plot for the apple shoot data.

Table 4.6 Bland's data for long and short apple shoots

	Long Shoots				Short Shoots		
DAY	n_i	$\bar{y}_i$	s_i	DAY	n_i	$\bar{y}_i$	s_i
0	5	10.20	0.83	0	5	10.00	0.00
3	5	10.40	0.54	6	5	11.00	0.72
7	5	10.60	0.54	9	5	10.00	0.72
13	6	12.50	0.83	19	11	13.36	1.03
18	5	12.00	1.41	27	7	14.29	0.95
24	4	15.00	0.82	30	8	14.50	1.19
25	6	15.17	0.76	32	8	15.38	0.51
32	5	17.00	0.72	34	5	16.60	0.89
38	7	18.71	0.74	36	6	15.50	0.54
42	9	19.22	0.84	38	7	16.86	1.35
44	10	20.00	1.26	40	4	17.50	0.58
49	19	20.32	1.00	42	3	17.33	1.52
52	14	22.07	1.20	44	8	18.00	0.76
55	11	22.64	1.76	48	22	18.46	0.75
58	9	22.78	0.84	50	7	17.71	0.95
61	14	23.93	1.16	55	24	19.42	0.78
69	10	25.50	0.98	58	15	20.60	0.62
73	12	25.08	1.94	61	12	21.00	0.73
76	9	26.67	1.23	64	15	22.33	0.89
88	7	28.00	1.01	67	10	22.20	0.79
100	10	31.67	1.42	75	14	23.86	1.09
106	7	32.14	2.28	79	12	24.42	1.00
				82	19	24.79	0.52
				85	5	25.00	1.01
				88	27	26.04	0.99
				91	5	26.60	0.54
				94	16	27.12	1.16
				97	12	26.83	0.59
				100	10	28.70	0.47
				106	15	29.13	1.74

Table 4.7 Weighted regression of $\bar{y}_i$ on DAY

Variable	Estimate	Standard Error	t-Value
Intercept	9.974	0.314	31.74
DAY	0.217	0.0054	40.71

$$\hat{\sigma}^2 = 3.720 \qquad R^2 = 0.988 \qquad \text{d.f.} = 20$$

Analysis of variance

Source	d.f.	SS	MS
Regression on DAY	1	6164.	6164.
Residual	20	74.39	3.720

stem units on that day. The apparent linearity of this plot should encourage us to try to fit the straight line

$$Y = \beta_0 + \beta_1 DAY + error \tag{4.18}$$

Should this model fit, we would have the interesting result that the observed rate of production of stem units, estimated as $\hat{\beta}_1$ per day, is constant.

For each sampled day, in Table 4.6, we have the following information: DAY, n_i = number of shoots sampled that day, $\bar{y}_i$ = average number of stem units in the n_i shoots, and s_i = standard deviation of the observed y, that is, $s_i = [\sum (y - \bar{y}_i)^2 / (n_i - 1)]^{1/2}$, where the sum is over the n_i units taken on the ith day. If we assume that the variance in the number of stem units per shoot is constant over the growing season, then the variance of $\bar{y}_i$ is $var(\bar{y}_i) = \sigma^2 / n_i$. Thus to estimate the regression of the $\bar{y}_i$ on DAY, weighted least squares must be used, with weights $w_i = 1/n_i$. The results of this regression are given in Table 4.7. The summary statistics indicate that this model explains most of the variability in number of stem units.

However, a pooled estimate of σ^2 independent of the model can be obtained from the s_i by the formula

$$\hat{\sigma}^2_{pooled} = \frac{\sum (n_i - 1)s_i^2}{\sum (n_i - 1)} = \frac{255.2}{167} = 1.528 \tag{4.19}$$

with $\sum (n_i - 1) = 167$ degrees of freedom.

It can be shown that, if a regression analysis were done on the original 189 shoots collected in this study rather than on the day means, then all results given in Table 4.7 would stay the same, except that the residual sum of squares in Table 4.7 would correspond to the sum of squares for lack of fit, and the sum of squares for pure error would be $\sum (n_i - 1)s_i^2$.

Thus an F-test for lack of fit is given by

$$F = \frac{MS(lof)}{MS(pe)} = \frac{3.720}{1.528} = 2.43$$

Since $F(0.01; 20, 167) = 1.63$, the p-value is less than 0.01, indicating that the straight-line model (4.18) does not appear to be adequate. However, the F-test with this many degrees of freedom is very powerful, and will detect very small deviations from the null hypothesis. Thus while the result here is statistically significant, it may not be scientifically significant, and for the purposes of describing the growth of apple shoots the fit of (4.18) may be adequate.

4.4   General F testing

Thus far we have encountered several situations that lead to computation of a statistic that has a nominal F distribution when a null hypothesis and normality hold. The theory for the F-tests is quite general. In the basic structure, a smaller model (null hypothesis) is compared to a larger model (alternative hypothesis), and the smaller model can be obtained from the larger by setting some parameters in the larger model equal to zero, equal to each other, or equal to some specific value. One example previously encountered is testing to see if the last q variables in a multiple regression are needed after fitting the first $p' - q$. In matrix notation, partition $\mathbf{X} = (\mathbf{X}_1, \mathbf{X}_2)$, where $\mathbf{X}_1$ is $n \times (p' - q)$, $\mathbf{X}_2$ is $n \times q$, and partition $\boldsymbol{\beta}^T = (\boldsymbol{\beta}_1^T, \boldsymbol{\beta}_2^T)$, where $\boldsymbol{\beta}_1$ is $(p' - q) \times 1$, $\boldsymbol{\beta}_2$ is $q \times 1$, so the two hypotheses, NH and AH, are

$$
\begin{aligned}
\text{NH:} \quad & \mathbf{Y} = \mathbf{X}_1 \boldsymbol{\beta}_1 + \mathbf{e} \\
\text{AH:} \quad & \mathbf{Y} = \mathbf{X}_1 \boldsymbol{\beta}_1 + \mathbf{X}_2 \boldsymbol{\beta}_2 + \mathbf{e}
\end{aligned}
\tag{4.20}
$$

The smaller model is obtained from the larger by setting $\boldsymbol{\beta}_2 = 0$.

To compute the F-test, both of the models must be fit to observed data. Under NH, find the residual sum of squares and its degrees of freedom RSS_{NH} and d.f.$_{\text{NH}}$. Similarly, under the alternative model, find RSS_{AH} and d.f.$_{\text{AH}}$. Clearly d.f.$_{\text{NH}} >$ d.f.$_{\text{AH}}$, since the alternative fits more parameters. Also, $RSS_{\text{NH}} - RSS_{\text{AH}} \geq 0$, since the fit of the AH must be at least as good as the fit of the NH. The F-test then gives evidence against NH if

$$
F = \frac{(RSS_{\text{NH}} - RSS_{\text{AH}})/(\text{d.f.}_{\text{NH}} - \text{d.f.}_{\text{AH}})}{RSS_{\text{AH}}/\text{d.f.}_{\text{AH}}}
\tag{4.21}
$$

is large when compared to the percentage points of $F(\text{d.f.}_{\text{NH}} - \text{d.f.}_{\text{AH}}, \text{d.f.}_{\text{AH}})$.

Non-null distributions. The numerator and denominator of (4.21) are independently distributed. Under normality and AH, each is distributed as σ^2 times a (noncentral) chi-squared variable. In particular, the expected value of the numerator of (4.21) will be

$$
E(\text{numerator of } (4.21)) = \sigma^2 + (\text{noncentrality parameter})^2 \tag{4.22}
$$

For the particular hypothesis of (4.20), the noncentrality parameter is given by the expression

$$
\boldsymbol{\beta}_2^T \left(\mathbf{X}_2^T \mathbf{X}_2 - \mathbf{X}_2^T \mathbf{X}_1 (\mathbf{X}_1^T \mathbf{X}_1)^{-1} \mathbf{X}_1^T \mathbf{X}_2 \right) \boldsymbol{\beta}_2 \tag{4.23}
$$

To help understand this, consider the special case of $\mathbf{X}_2^T \mathbf{X}_2 = I$ and $\mathbf{X}_1^T \mathbf{X}_2 = 0$ (i.e., the variables in $\mathbf{X}_2$ are orthogonal to each other and to the variables in X_1). Then (4.22) becomes

$$E(\text{numerator}) = \sigma^2 + \beta_2^T \beta_2 \tag{4.24}$$

Thus for this special case the expected value of the numerator of (4.21), and hence the power of the F-test, will be larger if β_2 is large. In the more general case where $X_1^T X_2 \neq 0$, the results are more complicated and the size of the noncentrality parameter, and hence power of the F-test depend not only on the size of β_2 but also on $X_1^T X_2$ or the *sample* correlations between the variables in X_1 and those in X_2. If these correlations are large then the power of F may be small even if β_2 is large.

More general results on F-tests are presented in advanced linear model texts such as Seber (1977).

Additional comments. The F-tests derived in this section have many important properties when the error terms are in fact normally distributed. For example they are likelihood ratio tests and all of the properties of such tests apply to the F-test. Since the exact normality of the error terms generally does not hold, discussion of optimality from a practical point of view is unnecessary. Fortunately, these procedures are "robust" to departures from normality of the errors; that is, estimates, tests, and confidence procedures are only modestly affected by departures from normality.

4.5   Joint confidence regions

Just as confidence intervals for a single parameter are based on the t distribution, confidence regions for several parameters will require use of an F distribution. These regions will be elliptically shaped.

The $(1 - \alpha) \times 100\%$ confidence region for β is the set of points β such that

$$\frac{(\beta - \hat{\beta})^T (X^T X)(\beta - \hat{\beta})}{p' \hat{\sigma}^2} \leq F(\alpha; p', n - p') \tag{4.25}$$

More typically, the confidence region for β^*, the parameter vector excluding β_0, is of interest. The $(1 - \alpha) \times 100\%$ region for β^* is, using the notation of Chapter 2, the set of points β^* such that

$$\frac{(\beta - \hat{\beta}^*)^T (\mathcal{X}^T \mathcal{X})(\beta^* - \hat{\beta}^*)}{p \hat{\sigma}^2} \leq F(\alpha; p, n - p') \tag{4.26}$$

The difference between (4.25) and (4.26) is the substitution of $\mathcal{X}^T \mathcal{X}$ for $X^T X$ and p for p'. The region (4.25) is a p'-dimensional ellipsoid centered at $\hat{\beta}$, while (4.26) is a p-dimensional ellipsoid centered at $\hat{\beta}^*$.

For example, the 95% confidence region for β_1, β_2 in the regression of FUEL on $X_1 = $ TAX and $X_2 = $ DLIC (Example 2.1) is given in Figure 4.3. This ellipse is centered at $(-32.075, 1251.46)$. The orientation of the ellipse (the directions of the major and minor axes) is determined by $\mathfrak{X}^T \mathfrak{X}$ or, equivalently, by the sample correlation between X_1 and X_2. If X_1 and X_2 are uncorrelated, the axes of the ellipse will be parallel to the X_1 and X_2 axes.

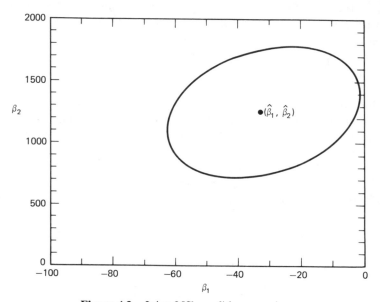

Figure 4.3   Joint 95% confidence region.

Confidence ellipsoid for an arbitrary subset of β. Rather than give the general results, only a special case will be given from which the general rule is to be derived. Suppose the 95% confidence region for $(\beta_1, \beta_2)^T$ is desired from a model with four variables in the fuel consumption data. Let **S** be the 2×2 submatrix of $(\mathbf{X}^T\mathbf{X})^{-1}$ corresponding to X_1 and X_2 (TAX and DLIC in the example). That is, from Table 2.3,

$$\mathbf{S} = \begin{bmatrix} 0.0382636 & 0.2215812 \\ 0.2215812 & 8.4108914 \end{bmatrix}$$

Then, the 95% confidence region is the set of points $\boldsymbol{\beta} = (\beta_1, \beta_2)^T$ such that

$$\frac{\left(\begin{pmatrix} \beta_1 \\ \beta_2 \end{pmatrix} - \begin{pmatrix} \hat{\beta}_1 \\ \hat{\beta}_2 \end{pmatrix} \right)^T \mathbf{S}^{-1} \left(\begin{pmatrix} \beta_1 \\ \beta_2 \end{pmatrix} - \begin{pmatrix} \hat{\beta}_1 \\ \hat{\beta}_2 \end{pmatrix} \right)}{2\hat{\sigma}^2} \leqslant F(\alpha; 2; n - p') \qquad (4.27)$$

where in (4.27), $(\hat{\beta}_1, \hat{\beta}_2)$ is computed from the four-variable model, $(\hat{\beta}_1, \hat{\beta}_2) = (-34.79, 1336.)$. This region is given in Figure 4.4, which for the example is not too different from the region in Figure 4.3. In other problems these regions may substantially differ, just as the parameter estimates may change, depending on the model.

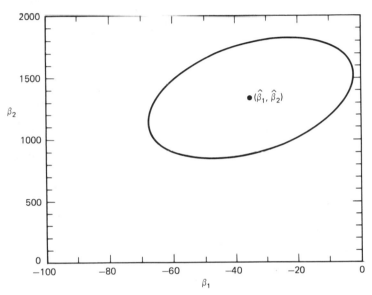

Figure 4.4 Joint conditional 95% confidence region.

Problems

4.1 Weights of sweet peas. The term regression was first used in connection with the work of Sir Francis Galton on inheritance of characteristics. In a paper on "Typical Laws of Heredity," delivered to the Royal Institution on February 9, 1877, Galton discussed some experiments he had instigated using sweet peas. By comparing the sweet peas produced by parent plants to those produced by offspring plants, he could observe the inheritance from generation to generation. One obvious characteristic of sweet peas is their diameter. To obtain data, Galton categorized plants according to the typical diameter of the peas they produced. For each of 7 size classes (15 to 21 hundredths of an inch), he arranged for each of 9 of his friends to grow 10 plants from seed in each size class; however, 2 of the crops were total failures. Galton's data were later published by Karl

Pearson (1914), as given in Table 4.8. In the table, only the mean diameters of the offspring seed are given along with respective standard deviations; sample sizes are unknown.

Table 4.8   Galton's data

Diameter of Parent Peas (hundredths of an inch)	Mean Diameter of Offspring Peas (hundredths of an inch)	Standard Deviation
21	17.26	1.988
20	17.07	1.938
19	16.37	1.896
18	16.40	2.037
17	16.13	1.654
16	16.17	1.594
15	15.98	1.763

4.1.1. Draw the scatter plot of $Y =$ mean offspring diameter versus $X =$ parent diameter.

4.1.2. Assuming that the standard deviations in Table 4.8 are exactly correct (i.e., population standard deviations), compute the weighted regression of Y on X. Draw the fitted line into your scatter plot.

4.1.3. Test the hypothesis that the slope parameter β_1 is equal to 1 versus the alternative that it is arbitrary.

4.1.4. Obtain a test for lack of fit of the straight-line model.

4.1.5. In his experiments, Galton took the average size of all peas produced by a plant to determine the size class of the parental plant. Yet for seeds to represent that plant and produce offspring, Galton chose seeds that were as close to the overall average size as possible. Thus for a small plant, the exceptional healthy seed was chosen as a representative, while robust plants were represented by relatively weaker seeds. What effects would you expect these experimental biases to have on (1) estimates of slope and intercept, and (2) estimates of error?

4.2   Physics data. Table 4.9 lists the results of the experiments like those described in Example 4.1 with choices of $a = \pi^+$ and π^- and $c = \pi^+$ and π^- from Weisberg et al. (1978). The data with $a = c = \pi^-$ are also given in Table 4.1. In Table 4.9 the values to the right of the $\pm$ sign are the values of $\sigma\sqrt{w_i}$; this is typical scientific notation.

4.2.1. For one or more of these data sets, fit model (4.13) and test for lack of fit. If the model does not fit, try model (4.17), and again test for lack of fit. Theoretically, for fixed c, the values of β_0 and β_1 for $a = \pi^-$ should equal those for $a = \pi^+$. Informally compare the fitted models for relevant combinations of a and c.

Table 4.9 Strong interaction data

p_a^{lab} (GeV/c)	$s^{-1/2}$ GeV/c^{-1}	$y(ap \to \pi^-)$ (μb)	$y(ap \to \pi^+)$ (μb)
	$a = \pi^-$		
4	0.345	367 ± 17	284 ± 13
6	0.287	311 ± 9	288 ± 9
8	0.251	295 ± 9	304 ± 9
10	0.225	268 ± 7	284 ± 8
12	0.207	253 ± 7	281 ± 7
15	0.186	239 ± 6	276 ± 7
20	0.161	220 ± 6	275 ± 7
24	0.148		257 ± 10
30	0.132	213 ± 6	275 ± 7
75	0.084	193 ± 5	277 ± 7
150	0.060	192 ± 5	281 ± 7
	$a = \pi^+$		
4	0.345	133 ± 6	499 ± 18
6	0.287	157 ± 6	464 ± 13
8	0.251	157 ± 5	442 ± 12
10	0.225	154 ± 6	389 ± 11
12	0.207	149 ± 6	388 ± 12
15	0.186	156 ± 10	341 ± 14
20	0.161	173 ± 22	375 ± 22
30	0.132	155 ± 5	331 ± 9
75	0.084	166 ± 5	303 ± 8
150	0.060	160 ± 6	311 ± 9
250	0.046	171 ± 11	306 ± 18

4.3 Apple shoots. Apply the analysis of Section 4.3 to the data on short shoots in Table 4.6. Informally compare the fitted regression for short and long shoots (be sure to draw a scatter plot of $\bar{y}_i$ vs. DAY).

4.4 Control charts. The data in Table 4.10 give the outside diameter of crankpins produced by an industrial process over several days (Jensen, 1977). All of the crankpins produced should be between 0.7425 and 0.7430 inches. The numbers given in the table are in units of 0.00001 inches deviation from 0.742 inches. For example, the number 93 means 0.742 + 0.00093 = 0.74293 inches.

When the manufacturing process is "under control," the average size of crankpins produced should (1) fall near the middle of the specified range and (2) should not depend on time. Fit the appropriate model to see if the process is under control, and test for lack of fit of the model.

Table 4.10   Crankpin data

Day	Diameter of Crankpins
1	93, 98, 90, 94, 94
4	93, 100, 88, 85, 89
7	89, 90, 92, 95, 100
10	93, 88, 87, 87, 87
13	88, 86, 91, 89, 86
16	82, 72, 80, 72, 89
19	81, 80, 78, 94, 90
22	90, 92, 82, 77, 89

4.5   An F-test. In simple regression derive an explicit formula for the F-test of

$$\text{NH:}\quad y_i = x_i + e_i \quad\quad (\text{e.g., } \beta_0 = 0, \beta_1 = 1)$$

$$\text{AH:}\quad y_i = \beta_0 + \beta_1 x_i + e_i \quad\quad i = 1, 2, \ldots, n$$

4.6   Snow geese. Aerial survey methods are regularly used to estimate the number of snow geese in their summer range areas west of Hudson Bay in Canada. To obtain estimates, small aircraft fly over the range and, when a flock of geese is spotted, an experienced person estimates the number of geese in the flock. To investigate the reliability of this method of counting, an experiment was conducted in which an airplane carrying two observers flew over $n = 45$ flocks, and each observer made an independent estimate of the number of birds in each flock. Also, a photograph of the flock was taken so that an exact count of the number of birds in the flock could be made. The resulting data are given in Table 4.11 (Cook and Jacobson, 1978).

4.6.1. Draw scatter plots of $Y = $ photo count versus $X_1 = $ count by observer 1 and versus $X_2 = $ count by observer 2. Do these graphs suggest that a simple regression model might be appropriate? Why or why not? For the simple regression model of Y on X_1 or on X_2, what do the error terms measure? Why is it appropriate to fit the regression of Y on X_1 or X_2 rather than the regression of X_1 or X_2 on Y?

4.6.2. Compute the regression of Y on X_1 and Y on X_2 via ordinary least squares, and test the hypothesis of Problem 4.5 for each observer. State in words the meaning of this hypothesis, and the result of the test. Are either of the observers reliable (you must define reliable)? Summarize your results.

Table 4.11 Estimates of flock size

Photo	Observer Number 1	Observer Number 2
56	50	40
38	25	30
25	30	40
48	35	45
38	25	30
22	20	20
22	12	20
42	34	35
34	20	30
14	10	12
30	25	30
9	10	10
18	15	18
25	20	30
62	40	50
26	30	20
88	75	120
56	35	60
11	9	10
66	55	80
42	30	35
30	25	30
90	40	120
119	75	200
165	100	200
152	150	150
205	120	200
409	250	300
342	500	500
200	200	300
73	50	40
123	75	80
150	150	120
70	50	60
90	60	100
110	75	120
95	150	150
57	40	40
43	25	35
55	100	110
325	200	400
114	60	120
83	40	40
91	35	60
56	20	40

4.6.3. Repeat 4.6.2, except fit the regression of $Y^{1/2}$ on $X_1^{1/2}$ and $Y^{1/2}$ on $X_2^{1/2}$. The square-root scale is used to stabilize the error variance.

4.6.4. Repeat 4.6.2, except assume that $\text{var}(e_i) = x_i \sigma^2$.

As a result of this experiment, the practice of using visual counts of flock size to determine population estimates was discontinued in favor of using photographs.

5

CASE ANALYSIS I:
RESIDUALS AND INFLUENCE

Thus far we have studied techniques that are used for the analysis of data when the multiple regression model fit is appropriate, and, as long as the model is known to be correct, no further analysis is required. However, the data analyst should almost never make assumptions without careful checks on them. With the exception of the discussion of Sections 4.3 and 4.4, where additional information is used, none of the methods for getting parameter estimates, testing, and the like can be used to obtain checks on the adequacy of a model. Rather, the data must be analyzed in detail, with attention given to the role of each case in determining values of estimators and test statistics. In this second phase of regression analysis, statistics will be computed that, rather than combining information from all the cases to produce a few aggregates, will have separate values for each case in the data. Consequently, we call these *case statistics*, and the whole technique is called *case analysis*.

The primary concerns of case analysis are two interrelated questions. First, we ask how well the model used resembles the data actually observed. The basic statistic here will be the residual (actually, a convenient transformation of it). If the fitted model does not give a set of residuals that seem reasonable, then some aspect of the model will be called into doubt. The second question of interest is the effect of each case on estimation and other aspects of aggregate analysis. In some data sets, for example, it may be that the observed aggregate statistics depend on one case in such a way that, if that case were deleted, the outcome of the aggregate analysis would change. We will term such a case influential, and shall learn ways of detecting and understanding influential cases. We will be led to study and use two case statistics that are relatively unfamiliar to many data analysts, called leverage values and distance measures.

Example 5.1   The usefulness of plots (Anscombe, 1973)

The need for case analysis is well illustrated by the four artificial data sets given in Table 5.1. Each set consists of 11 pairs of points (x_i, y_i), to which the simple linear regression model $y_i = \beta_0 + \beta_1 x_i + e_i$ is fit. Each data set leads to an identical aggregate analysis, namely,

$$\hat{\beta}_0 = 3.0$$
$$\hat{\beta}_1 = 0.5$$
$$\hat{\sigma}^2 = 13.75$$
$$R^2 = 0.667$$

Since the aggregate statistics are the same for each data set one might conclude that the linear regression model is equally appropriate for each of them. However, the simplest case analysis suggests that this is not true, as seen by examination of the scatter plots in Figures 5.1a through 5.1d.

Table 5.1   Four hypothetical data sets

	Data Set #					
	1–3	1	2	3	4	4
			Variable			
Case Number	X	Y	Y	Y	X	Y
1	10.0	8.04	9.14	7.46	8.0	6.58
2	8.0	6.95	8.14	6.77	8.0	5.76
3	13.0	7.58	8.74	12.74	8.0	7.71
4	9.0	8.81	8.77	7.11	8.0	8.84
5	11.0	8.33	9.26	7.81	8.0	8.47
6	14.0	9.96	8.10	8.84	8.0	7.04
7	6.0	7.24	6.13	6.08	8.0	5.25
8	4.0	4.26	3.10	5.39	19.0	12.50
9	12.0	10.84	9.13	8.15	8.0	5.56
10	7.0	4.82	7.26	6.42	8.0	7.91
11	5.0	5.68	4.74	5.73	8.0	6.89

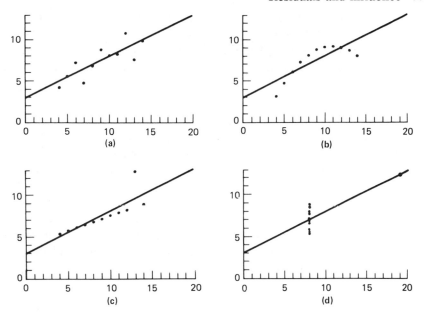

Figure 5.1 Four hypothetical data sets. Reproduced with permission from Anscombe (1973).

The first data set, graphed in Figure 5.1*a*, is as one might expect to observe if the simple linear regression model were appropriate. The graph of the second data set given in Figure 5.1*b* suggests a different conclusion, that the analysis based on simple linear regression is incorrect, and that a smooth curve, perhaps a quadratic polynomial, could be fit to the data with little remaining variability.

Figure 5.1*c* suggests that the prescription of simple regression may be correct for most of the data, but one of the cases is too far away from the fitted regression line. This is called the outlier problem. Possibly the case that does not match the others should be deleted from the data set, and the regression should be refitted from the remaining 10 cases. If this is done, the fitted equation is given by $\hat{y} = 4. + 0.346x$, which is quite different from that obtained from all 11 cases. Without a context for the data, we cannot judge one line "correct" and the other "incorrect." The differences between these two lines need to be understood and reported.

The final set, graphed in Figure 5.1*d*, is different from the other three in that there is really not enough information to make a judgment concerning the fitted model. The slope parameter estimate $\hat{\beta}_1$ is largely determined by the value of y_8. If the eighth case were deleted, we could not even estimate β_1. We must distrust an aggregate analysis that is so heavily dependent upon a single case.

For the case statistics to be useful, we must understand their behavior when assumptions about the model are true, and also we must know exactly how specific failures in assumptions will effect them. With this knowledge, the case statistics can then be used to diagnose the success or failure of the model with respect to observed data. Their behavior when the model is correct is the main topic of this chapter. First, residuals, their properties, and useful scaling of residuals for case analysis are discussed. Then, we turn to the *empirical influence function* that will lead us to consider statistics that directly measure the impact of each case on regression. Finally, we turn to the problem of detecting outliers, and begin to answer the questions of what to do about them.

5.1   The residuals

In studying the residuals, it is convenient to use matrix notation, as outlined in Chapter 2. The basic model that we consider is

$$\mathbf{Y} = \mathbf{X}\boldsymbol{\beta} + \mathbf{e} \qquad \text{var}(\mathbf{e}) = \sigma^2 \mathbf{I}_n \tag{5.1}$$

In (5.1), we assume that $\mathbf{X}$ is a known full rank matrix of n rows and p' columns, where $p' = $ number of independent variables $+ 1 = p + 1$ if a column of ones is included in the model, and $p' = p$ if the column of ones is not in the model. Similarly, $\boldsymbol{\beta}$ is a $p' \times 1$ unknown vector. The vector $\mathbf{e}$ consists of unknown errors which we assume are equally variable and uncorrelated.

In fitting the model (5.1), we first estimate $\boldsymbol{\beta}$ by $\hat{\boldsymbol{\beta}} = (\mathbf{X}^T\mathbf{X})^{-1}\mathbf{X}^T\mathbf{Y}$, and the fitted values $\hat{\mathbf{Y}}$ corresponding to the observed values $\mathbf{Y}$ are then given by

$$\hat{\mathbf{Y}} = \mathbf{X}\hat{\boldsymbol{\beta}}$$

$$= \mathbf{X}\left[(\mathbf{X}^T\mathbf{X})^{-1}\mathbf{X}^T\mathbf{Y}\right]$$

$$= \mathbf{X}(\mathbf{X}^T\mathbf{X})^{-1}\mathbf{X}^T\mathbf{Y} \tag{5.2}$$

The vector of residuals $\hat{e}$ is defined by

$$\hat{e} = Y - \hat{Y}$$
$$= Y - X(X^TX)^{-1}X^TY$$
$$= \left[I - X(X^TX)^{-1}X^T\right]Y \tag{5.3}$$

The matrix defined by $X(X^TX)^{-1}X^T$ is very important in our study, so we shall give it a name, and define V by

$$V = X(X^TX)^{-1}X^T \tag{5.4}$$

Using this definition, the fitted values are given by $\hat{Y} = VY$ and the residuals are given by $\hat{e} = (I - V)Y$. In fact, we may write, in analogy to (5.1), the relationship

$$Y = (V + I - V)Y$$
$$= VY + (I - V)Y$$
$$= X\hat{\beta} + \hat{e} \tag{5.5}$$

Careful comparison of (5.5) with (5.1) reminds us that the vector of residuals $\hat{e}$ is not the same as the vector of errors e. To examine the appropriateness of (5.1), we would like to have e available for study. But, since we only actually get $\hat{e}$, we can study only whether or not the fitted model (5.5) matches the data.

Differences between e and ê. By assumption, the errors are uncorrelated random variables with zero means, common variance σ^2. The moments of $\hat{e}$ are found using (5.3) and the results in Appendix 2A.2,

$$E(\hat{e}) = 0$$
$$V(\hat{e}) = \sigma^2(I - V) \tag{5.6}$$

Like the errors, each of the residuals has zero mean, but they all have different variances and they are all correlated.

If the errors are normally distributed, then so are the residuals, since, from (5.3) the residuals are just a linear combination of the elements of Y or of e. Also, if the constant term (the intercept) is included in the model, then for least squares fitting the sum of the residuals is zero, $\sum \hat{e}_i = \hat{e}^T 1 = 0$.

The elements of V, the v_{ij}'s, are given by the equation

$$v_{ij} = x_i^T(X^TX)^{-1}x_j \tag{5.7}$$

and, for the diagonal elements,

$$v_{ii} = \mathbf{x}_i^T (\mathbf{X}^T \mathbf{X})^{-1} \mathbf{x}_i \qquad (5.8)$$

where $\mathbf{x}_i^T$ and $\mathbf{x}_j^T$ are, respectively, the ith row and the jth row of the data matrix $\mathbf{X}$. From (5.6), we see that the variance of the ith residual is

$$\text{var}(\hat{e}_i) = \sigma^2 (1 - v_{ii}) \qquad (5.9)$$

and the covariance between the ith and the jth residual is

$$\text{cov}(\hat{e}_i, \hat{e}_j) = -\sigma^2 v_{ij} \qquad (5.10)$$

Also, the correlation between the ith and the jth residual is

$$\text{corr}(\hat{e}_i, \hat{e}_j) = -v_{ij} / (1 - v_{ii})^{1/2} (1 - v_{jj})^{1/2} \qquad (5.11)$$

Now, $\mathbf{V}$ is an $n \times n$ symmetric matrix. In addition, since $\mathbf{V}^2 = \mathbf{V}\mathbf{V} = \mathbf{V}$ (as verified by direct multiplication), $\mathbf{V}$ is called idempotent. Idempotent matrices play a central role in least squares theory, and much is known about them (Seber, 1977; Graybill, 1969). In particular, $\mathbf{V}$ will generally be of less than full rank, and $\text{rank}(\mathbf{V}) = \text{rank}(\mathbf{X}) = p'$. Also, the sum of the diagonal elements, called its trace, is

$$\sum_{i=1}^{n} v_{ii} = \text{rank}(\mathbf{X}) = p' \qquad (5.12)$$

Finally, $\sum_j v_{ij}^2 = v_{ii}$ (verified directly from the definition of idempotency), and, if the constant is in the model,

$$\sum_i v_{ij} = \sum_j v_{ij} = 1 \qquad (5.13)$$

It is clear that each v_{ii} must fall in the range between 0 and 1. Actually, tighter bounds on v_{ii} are possible. If we let c be the number of rows in $\mathbf{X}$ that are exactly the same as $\mathbf{x}_i$, including $\mathbf{x}_i$, then, for models with an intercept term,

$$\frac{1}{n} \leqslant v_{ii} \leqslant \frac{1}{c} \qquad (5.14)$$

For example, if there are two identical rows in $\mathbf{X}$, the corresponding v_{ii} are bounded between $1/n$ and $\frac{1}{2}$. The lower bound for v_{ii} is attained only if $\mathbf{x}_i$ is exactly equal to the vector of sample averages.

Example 5.2   Simple regression

Consider now the simple regression model of Chapter 1, in which $\hat{y}_i = \hat{\beta}_0 + \hat{\beta}_1 x_i$. The matrix $\mathbf{V}$ is $n \times n$, and even for simple regression the number of entries in $\mathbf{V}$, namely, n^2, may be quite large, and it is rarely computed in full. However, using (2.25) to get $(\mathbf{X}^T\mathbf{X})^{-1}$, a formula may be obtained for an individual v_{ij}. We find

$$v_{ij} = \mathbf{x}_i^T(\mathbf{X}^T\mathbf{X})^{-1}\mathbf{x}_j$$

$$= (1 \;\; x_i)\begin{bmatrix} \dfrac{\sum x_i^2}{nSXX} & \dfrac{-\bar{x}}{SXX} \\[2mm] \dfrac{-\bar{x}}{SXX} & \dfrac{1}{SXX} \end{bmatrix}\begin{bmatrix} 1 \\[2mm] x_j \end{bmatrix}$$

$$= \frac{1}{n} + \frac{(x_i - \bar{x})\,(x_j - \bar{x})}{SXX} \tag{5.15}$$

By setting j equal to i in (5.15) we find the diagonal elements v_{ii} to be

$$v_{ii} = \frac{1}{n} + \frac{(x_i - \bar{x})^2}{SXX} \tag{5.16}$$

Note that $\sum_{i=1}^{n} v_{ii} = 2$, as mentioned above or by direct summation in (5.15). Now, v_{ii} will be minimized ($v_{ii} = 1/n$) if $x_i = \bar{x}$, and v_{ii} will increase for "unusual" x_i, that is, for x_i far from $\bar{x}$. If x_i is sufficiently far from $\bar{x}$, the term $(x_i - \bar{x})^2/SXX$ will become larger and v_{ii} may be arbitrarily close to one.

The notion of v_{ii} giving a measure of how far the ith case is from the center of the data is central to case analysis. As can be seen from (5.9), cases with large values of v_{ii} will have small values for var($\hat{e}_i$); as v_{ii} gets closer to one, this variance will approach zero, and, as this happens, no matter what value of y_i is observed for the ith case, we are nearly certain to get a residual for the ith case near zero; that is, $\hat{y}_i \cong y_i$. Such a case can be very important in estimating parameters. Cases with large v_{ii} have been called high leverage cases (Hoaglin and Welsch, 1978). We would like to be able to characterize those cases that will have large v_{ii} and those that have small v_{ii}.

In multiple regression, v_{ii} measures the distance from the point $\mathbf{x}_i$ to the center of the data, and cases with unusual values for the independent variables will tend to have large values of v_{ii}. There are many ways that we can formalize the distance notion. The one that works (at least for models with an intercept) is best defined in terms of the deviations from means form of the model discussed in Chapter 2. Suppose that $\mathcal{X}^T\mathcal{X}$ is the corrected cross-product matrix (2.19), $\bar{\mathbf{x}}$ is the $p \times 1$ vector of sample means of the p-independent variables, and we redefine $\mathbf{x}_i^T$ to be the ith row of $\mathbf{X}$ *without* the one for the intercept. Then, one can show that v_{ii} can be written as

$$v_{ii} = \frac{1}{n} + (\mathbf{x}_i - \bar{\mathbf{x}})^T (\mathfrak{X}^T \mathfrak{X})^{-1} (\mathbf{x}_i - \bar{\mathbf{x}}) \tag{5.17}$$

Geometrically, the second term on the right-hand side of (5.17) gives the equation of an ellipsoid centered at $\bar{\mathbf{x}}$.

For example, consider again the data on fuel consumption first discussed in Chapter 2. We will only use the model with $X_1 = $ TAX and $X_2 = $ DLIC as predictors so that a two-dimensional picture is possible. The data for (X_1, X_2) are given in the scatter plot in Figure 5.2. The ellipses drawn on the graph correspond to elliptical contours of constant v_{ii} for v_{ii} between 0.25 and 0.05. Thus, for example, any point that falls exactly on the outer contour would have $v_{ii} = 0.25$, while points on the innermost contour have $v_{ii} = 10.05$. This definition of distance depends on the data. In fact, points near the long or major axis of the ellipsoid need to be much farther away from $\bar{x}$, in the usual Euclidean distance sense, than do points closer to the minor axis, to have the same values for v_{ii}.

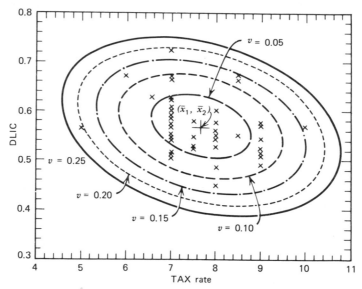

Figure 5.2   Contours of constant v_{ii} in two dimensions.

In the example, the state with the lowest taxes—Texas, with 5¢ tax—has the largest v_{ii}, about 0.2. None of the v_{ii} are very large, although in other data sets finding cases with v_{ii} between 0.5 and 1.0 is not unusual. Cases with large v_{ii} are potentially the most important in fitting a model. Note again that this potential influence does not depend on Y, but only on the X's. Comprehensive measures of influence must consider both of these.

The Mahalanobis distance. If the $1/n$ term in (5.17) is dropped from the right-hand side of the equation, and the remaining term is multiplied by $(n - 1)$, the resulting quantity is called the Mahalanobis distance from x_i to the center of the data $\bar{x}$. The Mahalanobis distance is widely used in multivariate analysis, most notably in discriminant analysis, as a measure for assigning units to different populations based on their relative location. The use of the Mahalanobis distance in discriminant analysis is founded on multivariate normality of the rows of X, an assumption we neither need nor use.

Studentized residuals. As already pointed out, var($\hat{e}_i$) will be small whenever v_{ii} is large, so cases with x_i near $\bar{x}$ will be fit poorly, and cases with x_i far from $\bar{x}$ will be fit well. This is particularly undesirable because violations of a model may be most likely to occur under unusual conditions and we may be unable to find those violations, at least by simply examining the residuals, as those residuals tend to be smaller.

The discussion of the last paragraph suggests that an improved set of residuals can be obtained by scaling so that cases with large v_{ii} get larger scaled residuals, and cases with smaller v_{ii} get smaller scaled residuals. Then, all of the residuals in the analysis can be compared directly. One good way of doing this scaling is to divide each of the residuals by an estimate of its standard deviation. We call these the *Studentized residuals*, for which we use the symbol r_i, defined by

$$r_i = \frac{\hat{e}_i}{\hat{\sigma}\sqrt{1 - v_{ii}}} \qquad i = 1, 2, \ldots, n \qquad (5.18)$$

Unlike the $\hat{e}_i$, the sum of the r_i's is not zero, although $E(r_i) = 0$, $i = 1, 2, \ldots, n$. Also, the r_i's are slightly correlated with the $\hat{y}_i$'s, although in practice this correlation is negligible. The most important advantage of the Studentized residuals is that var(r_i) = 1 for all i, independent of σ^2 and the v_{ii}'s, as long as the model is correct; when the model is incorrect, then var(r_i) will generally not be constant over all i. The covariance of r_i and r_j is the correlation between the $\hat{e}_i$ and $\hat{e}_j$, e.g.,

$$\text{cov}(r_i, r_j) = - \frac{- v_{ij}}{(1 - v_{ii})^{1/2}(1 - v_{jj})^{1/2}}$$

Also, the distribution of r_i will be shown in Section 5.4 to be a monotonic transformation of a Student's t-distribution, and this statistic will in turn be appropriate for testing for outliers. As given, r_i is not distributed as t because the numerator and the denominator are not independent. As a first approximation, however, one may treat the r_i's as if they were all distributed as standard normal random variables. Thus the examination of residuals is "standardized" in the sense that the residual variability is nearly always the same when the model is true.

5.2   The influence function

The residuals, or the Studentized residuals just introduced, are the most commonly used case statistics. They try to measure the success or failure of fitting a model at each case. Now, a related question may be asked concerning the impact of the values observed for the ith case upon the values of the estimates. In some data sets, one of the cases may have sufficient impact upon the regression such that, if the case were not in the data, completely different results would have been obtained. This does not necessarily mean that the results of the regression with that case included are invalid; however, whether or not valid, it is important for the analyst to be able to recognize influential cases and then, within the context of a specific problem, try to make intelligent decisions as to what should be done about them.

The idea of measuring the influence of each case on the regression leads us to look at the data in a slightly different way. First, delete the ith case from the data, so that only $n - 1$ cases are left for analysis. Then, using this reduced data, recompute the aggregate statistics. Examine the difference between the estimates of parameters with the ith case included and with the ith case deleted; cases that lead to estimates that are quite different will be called influential. Although this procedure sounds tedious, as it appears that $n + 1$ regressions must be done, all of the necessary quantities can be computed from the r_i, the v_{ii}, $\hat{\sigma}^2$, n, and p'.

The empirical influence function. The influence function is a very useful mathematical construction for studying the behavior of estimates. Defined in terms of population parameters, it monitors the change in estimates when the data used are slightly modified. Its original use was to compare different methods of estimation, and so-called robust methods are defined in such a way that the influence function has good properties. (See Andrews, et al. (1972), Hinkley (1977), and Hampel (1974) for more complete discussions of influence functions.)

In the context of case analysis, rather than comparing estimation techniques, we wish to study the change in the estimate of β when a case is deleted from the data. We view the estimate $\hat{\beta}$ from the full sample, as a fixed point, and let $\hat{\beta}_{-i}$ be the least squares estimate of β obtained from the regression using all of the cases *except* the ith. An empirical version of the influence function that is very useful for our purposes is obtained by taking the difference between the full data estimate and the estimate using $n - 1$ cases (excluding the ith), $\hat{\beta}_{-i} - \hat{\beta}$. We will call a case influential if deleting the ith case results in a substantial change in the estimate of the parameter vector.

For example, using the fuel consumption data, again with $X_1 = \text{TAX}$ and $X_2 = \text{DLIC}$ as predictors, Figure 5.3 is a graph of $\hat{\beta}_{1, -i}$ versus $\hat{\beta}_{2, -i}$. For example, the point labeled "Wyoming" corresponds to the estimate of the slopes for TAX and DLIC that would be obtained if Wyoming was deleted from the data set (but all other cases were left the same). (To display completely the effect of deleting a case a three-dimensional graph is required, as the values of the intercept $\hat{\beta}_{0, -i}$ are not shown.)

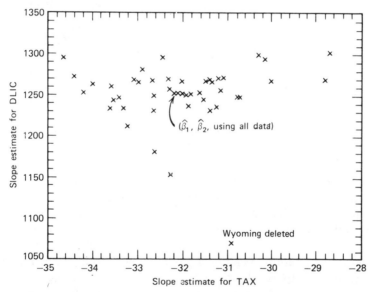

Figure 5.3   Estimates of (β_1, β_2) obtained by deleting each case in turn.

As is evident from Figure 5.3, deleting a case will result in changing the estimates; however, for most of the cases the new estimates will not be very far from the estimate based on all the data. For example, the estimates of β_1 and β_2 in the full data are $(-32.075, 1251.5)$; if Texas is deleted, the estimates are $(-33.549, 1243.6)$, and if only Wyoming is deleted the estimates are $(-30.933, 1069.1)$. From the graph this latter point appears to be relatively far removed from the original estimate. A method of measuring the distance between these points is needed so that we can judge if the case—here, Wyoming—has sufficient influence on the estimation of parameters that deletion of it would result in substantially different conclusions.

Cook's distance. To compute a distance between $\hat{\beta}$ and $\hat{\beta}_{-i}$ we appeal to the ideas used in Section 4.5 to obtain a confidence region for $\hat{\beta}$. In analogy to (4.25) and following Cook (1977, 1979), we define the (squared) distance from $\hat{\beta}_{-i}$ to $\hat{\beta}$ to be D_i, where

$$D_i = \frac{(\hat{\beta}_{-i} - \hat{\beta})^T (\mathbf{X}^T\mathbf{X})(\hat{\beta}_{-i} - \hat{\beta})}{p'\hat{\sigma}^2} \tag{5.19}$$

Defining distance in this way corresponds to making contours of constant D_i be ellipsoids, with the same structure as the confidence ellipsoids of Section 4.5. Cases with large values of D_i would have a substantial influence on estimation of β, and deletion of them may result in substantial alteration of conclusions.

By carrying the analogy to confidence ellipsoids one step further, suppose that an observed value of D_i was exactly equal to $F(a; p', n - p')$. This would mean that deleting case i would result in moving the estimate of β to the edge of a $(1 - a) \times 100\%$ confidence region based on the complete data. In the fuel example, for Wyoming (case 40), D_{40} can be shown to equal 0.3826. From a table of the F distribution with $p' = 3$, $n - p' = 48 - 3 = 45$ degrees of freedom, $F(0.9; 3, 45) = 0.194$ and $F(0.75; 3, 45) = 0.405$, so deleting Wyoming would move the estimate of β to the edge of about a 20% confidence region—a relatively small movement. Typically, cases with observed values of D_i larger than about 1, corresponding to movement to the edge of a 50% confidence region or beyond may be judged as influential cases. Note that D_i is *not* distributed as an F statistic; rather the comparison of D_i to F allows transformation of D_i to a familiar scale.

It is tempting to add the elliptical contours of constant D_i to Figure 5.3. However, the contours of constant D_i refer to the three-dimensional ellipsoid in the $(\beta_0, \beta_1, \beta_2)$ space, not to the two-dimensional space displayed in the graph (but see Problem 5.9 for a slightly different distance measure appropriate to Figure 5.3).

Computing D_i. From the derivation of D_i it is not clear that using these statistics is computationally feasible. However, the results sketched in Appendix 5A.1 are used to rewrite D_i using more familiar quantities. The best computational form for D_i is

$$D_i = \frac{1}{p'} r_i^2 \left(\frac{v_{ii}}{1 - v_{ii}} \right) \tag{5.20}$$

Thus aside from the constant p', D_i is a product of the square of the ith Studentized residual r_i (defined at (5.18)), and the ratio $v_{ii}/(1 - v_{ii})$ which can be shown to be the distance from x_i to the vector of sample means based on all of the data *except* the ith case, $0 \leqslant v_{ii}/(1 - v_{ii}) \leqslant \infty$. For fixed p', the size of D_i will be determined by two different sources: the magnitude of r_i, a random variable reflecting the lack of fit of the model at the ith case, and the distance of the vector x_i from the average of the other data vectors as reflected by v_{ii}. A large D_i may be determined by large r_i or by large leverage or both.

Table 5.2 gives the case statistics for three of the states in the fuel consumption data (again using only TAX and DLIC as predictors). To compute r_i and D_i from $\hat{e}_i$ and v_{ii}, we also need to know $\hat{\sigma}^2 = 5796.31$. For Wyoming, for example, we can compute

$$r_i = \frac{242.6}{\sqrt{5796.31(1 - 0.0930)}} = 3.3453$$

$$D_i = \frac{(3.3453)^2}{3} \frac{0.0930}{1 - 0.0930} = 0.3826$$

Table 5.2   Selected case statistics for fuel consumption data

State	$\hat{e}_i$	v_{ii}	r_i	D_i
18 ND	153.8	0.0444	2.0659	0.0661
37 TX	− 16.94	0.2067	− 0.2497	0.0054
40 WY	242.6	0.0930	3.3453	0.3826

We note that Texas, the state with the largest potential influence (largest v_{ii}) has relatively small observed influence (D_i small) because its Studentized residual r_i is very small. Similarly, Wyoming is relatively influential because of its large value of r_i—it appears that people in Wyoming consume more fuel than would be predicted from DLIC and TAX. However, in this example, none of the cases alone will exert substantial influence on the estimated parameters.

A complete case analysis requires consideration of the D_i, r_i, and v_{ii}, or equivalent functions of them, for each case.

Other measures of influence. Cook's distance provides a specific methodology for looking at the change in the estimate of β when a single case is deleted relative to the usual confidence ellipsoids. Clearly, other measures of influence are possible, either by concentrating on different aspects of the analysis, such as estimating a subset of β, or looking at groups of cases rather than just one at a time, or by altering the method of defining the measure. While these modifications may prove to be important in specific problems, the approach used here should be adequate for most purposes. A review of some of these modifications is given by Cook and Weisberg (1980).

Example 5.3   Rat data (Dennis Cook)

An experiment was conducted to investigate the amount of a particular drug present in the liver of a rat. Nineteen rats were randomly selected, weighed, placed under light ether anesthesia and given an oral dose of the drug. Because it was felt that large livers would absorb more of a given dose than smaller livers, the actual dose an animal received was approximately determined as 40 mg of the drug per kilogram of body weight. (Liver weight is known to be strongly related to body weight.) After a fixed length of time each rat was sacrificed, the liver weighed, and the percent of the dose in the liver determined.

The experimental hypothesis was that, for the method of determining the dose, there is no relationship between the percentage of the dose in the liver (Y) and the body weight (X_1), liver weight (X_2), and relative dose (X_3).

The data and sample correlations are given in Tables 5.3 and 5.4. As had been expected, the sample correlations between the response and the independent variables are all small, and none of the simple regressions of dose on any of the independent variables is significant, all having t-values less than 1 as shown in Table 5.5. However, the regression of Y on X_1, X_2, and X_3 gives a different and contradictory result: two of the independent variables, X_1 and X_3, have significant t-tests, with $p < 0.05$ in both cases, indicating that the two measurements combined are a useful indicator of Y; if X_2 is dropped from the model, the same phenomenon appears. The analysis so far, based only on aggregate statistics, might lead to the conclusion that a combination of dose and rat weight is associated with the response; however, dose was (approximately) determined to be a multiple of rat weight so, at least to a first approximation, rat weight and dose were measuring the same thing!

We turn to case analysis to attempt to resolve this paradox. In Table 5.6 the residuals and related statistics are listed for the model Y on X_1, X_2, X_3. The residuals ($\hat{e}_i$ or r_i) do not display any unusual features or reasons for the paradox: the $|r_i|$, for example, are all less than 2, without obvious trends or patterns. However, the D_i immediately locates a possible cause: case number 3 has $D_3 = 0.930$; no

Table 5.3 Rat data

X_1 = Body Weight (g)	X_2 = Liver Weight (g)	X_3 = Relative dose	Y
176	6.5	0.88	0.42
176	9.5	0.88	0.25
190	9.0	1.00	0.56
176	8.9	0.88	0.23
200	7.2	1.00	0.23
167	8.9	0.83	0.32
188	8.0	0.94	0.37
195	10.0	0.98	0.41
176	8.0	0.88	0.33
165	7.9	0.84	0.38
158	6.9	0.80	0.27
148	7.3	0.74	0.36
149	5.2	0.75	0.21
163	8.4	0.81	0.28
170	7.2	0.85	0.34
186	6.8	0.94	0.28
146	7.3	0.73	0.30
181	9.0	0.90	0.37
149	6.4	0.75	0.46

Table 5.4 Sample correlations—Rat data

X_1 = Body weight (g)	1.000			
X_2 = Liver weight (g)	0.500	1.000		
X_3 = Relative dose	0.990	0.490	1.000	
Y	0.151	0.203	0.228	1.000
	Body Weight	Liver Weight	Dose	Y

Table 5.5 Regression of Y on various predictors (t-values in parentheses)

	Model Including			
Coefficient	X_1	X_2	X_3	(X_1, X_2, X_3)
Intercept	0.196	0.220	0.133	0.266
	(0.89)	(1.64)	(0.63)	(1.37)
β_1 (rat weight)	0.0008			−0.0212
	(0.63)			(−2.67)
β_2 (liver weight)		0.0147		0.0143
		(0.86)		(0.83)
β_3 (dose)			0.235	4.178
			(0.96)	(2.74)

other case has D_i bigger than 0.273, suggesting that case number 3 alone may have large enough influence on the equation to induce the anomaly. The value of $v_{33} = 0.8509$ indicates that the problem with this case is that the vector $\mathbf{x}_3$ is different from the others.

One suggestion at this point is to delete the third case and recompute the regression. These computations are given in Table 5.7. Here, the paradox dissolves, and the apparent relationship found in the first analysis can thus be ascribed to the third case alone.

Table 5.6   Residuals from Y on X_1, X_2, X_3, all with $n = 19$ cases

Case Number	y_i	$\hat{e}_i$	r_i	v_{ii}	D_i
1	0.4200	0.1238	1.7660	0.1780	0.1688
2	0.2500	$-0.8914E-01$	-1.2730	0.1793	0.0885
3	0.5600	$0.2409E-01$	0.8072	0.8509	0.9296
4	0.2300	-0.1006	-1.3772	0.1076	0.0572
5	0.2300	$-0.6771E-01$	-1.1231	0.3915	0.2029
6	0.3200	$0.7131E-02$	0.1007	0.1612	0.0005
7	0.3700	$0.5658E-01$	0.7880	0.1369	0.0246
8	0.4100	$0.4958E-01$	0.7426	0.2537	0.0469
9	0.3300	$0.1231E-01$	0.1649	0.0670	0.0005
10	0.3800	$-0.2845E-02$	-0.0392	0.1197	0.0001
11	0.2700	$-0.8015E-01$	-1.1051	0.1195	0.0414
12	0.3600	$0.4236E-01$	0.6024	0.1724	0.0189
13	0.2100	$-0.9815E-01$	-1.5357	0.3162	0.2726
14	0.2800	$-0.2714E-01$	-0.3768	0.1314	0.0054
15	0.3400	$0.3161E-01$	0.4256	0.0762	0.0037
16	0.2800	$-0.5875E-01$	-0.8589	0.2166	0.0510
17	0.3000	$-0.1835E-01$	-0.2647	0.1952	0.0042
18	0.3700	$0.6068E-01$	0.8509	0.1487	0.0316
19	0.4600	0.1347	1.9220	0.1780	0.1999

Table 5.7   Regression with case 3 deleted

	Estimate	Standard Error	t-Value
Intercept	0.311	0.205	1.52
X_1	$-0.788E-02$	$0.187E-01$	-0.42
X_2	$0.899E-02$	$0.187E-01$	0.48
X_3	1.485	3.713	0.40

$\hat{\sigma}^2 = 0.612E-02$, $R^2 = 0.0211$, d.f. $= 14$

The careful analyst must now try to understand exactly why the third case is so influential. Inspection of the data indicates that this rat, with weight 190 g, was reported to have received a full dose of 1.000, which was a larger dose than it should have received according to the rule for assigning doses (for example, rat 8 with weight of 195 g got a lower dose of 0.98). A number of causes for the result found in the first analysis are possible: (1) the dose or weight recorded for case 3 was in error, so the case should probably be deleted from the study, or (2) the regression fit in the second analysis is not appropriate except in the region defined by the 18 points excluding case 3. This has many implications concerning the experiment. It is possible that the combination of dose and rat weight chosen was fortuitous, and that the lack of relationship found would not persist for any other combinations of them, since inclusion of a data point apparently taken under different conditions leads to a different conclusion. This suggests the need for collection of additional data, with dose determined by some rule other than a constant proportion of weight.

5.3 Outliers

It is not unusual in applications of regression for there to be one or more cases in a case analysis with observed Y values that fail to conform to a model, while the bulk of the data appear to correspond quite well. In the simple regression case (e.g., Figure 5.1c) this may be obvious from a plot of y_i against x_i, if most of the cases lie near a fitted line but one (or more) do not. If a case really does not fit in with the remaining ones, there are several possibilities.

1. An improbable but perfectly conforming observation was made. That is, y_i satisfies the fitted linear model, but the e_i associated with y_i happened to be large, as random deviates will occasionally be. A case like this should be included in the estimation process, and any procedure for handling outliers must protect the analyst against regularly discarding cases of this type.

2. After careful scrutiny of the conditions under which the data were collected, the case is found to correspond to exceptional though explainable circumstances, such as a power failure, failure of a measuring instrument, or the first day of a newly hired technician. One is probably justified in eliminating the case from the data set and estimating the regression model without it, since the process attached to this case is probably different from that for the rest of the data.

3. An exceptional event in fact happened as in (2), but no specific reason for it can be found. Here again, one would probably wish to delete the case from the analysis when estimating parameters.

4. The case is perfectly legitimate; nothing exceptional or even improbable occurred. However, the model for Y for this combination of X-values does not conform to the line or plane that describes (most of) the cases. This case may then be the most important in a study, as it could represent new and unexpected information. The researcher may wish to study the conditions of this point separately. However, a linear model may still be appropriate for the rest of the data so that analysis might often proceed by deleting this case.

In (2), (3), and (4) the conclusion is the same for the estimation of regression coefficients: compute the regression excluding the case that does not belong. Even when it is possible to identify "outlying" cases one can rarely distinguish between (3) and (4). However, unbiased estimation in (2), (3), and (4) does require identification of outliers. Since (1) is always a possibility, there is always a chance that if we exclude any case we are discarding real information. Moreover, because outlier rules usually discard cases with large residuals, there is an overall tendency to underestimate error variance. We need solutions to three problems: identification of possible outliers, assessment of the effect of outliers on the estimates, and protection against discarding good data. In practice, the last is a requirement for a test of the hypothesis that the suspected case does not conform to the same model as the remaining points. This test must take into account the selection problem inherent on basing a test on the largest of many statistics. For this reason, even though the test we use is nominally distributed as Student's t, special tables of critical values will be required.

In general, an outlier is a case that is inconsistent with the rest of the data. This vague definition is required to cover the variety of problems in which the term is applied. For our purposes here, however, a very specific model for outliers will be used, although others that are equally plausible are possible. A more general treatment of outliers is given by Barnett and Lewis (1978). Here, an outlier will be a case whose expectation $E(y_i)$ is very different from $\mathbf{x}_i^T\boldsymbol{\beta}$, while for the remaining data $E(y_{i'}) \cong \mathbf{x}_{i'}^T\boldsymbol{\beta}$, $i' \neq i$. This suggests that, for outliers, the residual $\hat{e}_i = y_i - \mathbf{x}_i^T\hat{\boldsymbol{\beta}}$, which is the sample analog of the difference $E(y_i) - \mathbf{x}_i^T\boldsymbol{\beta}$, will tend to be different from zero, and the best candidates for outliers are cases with large values of $|\hat{e}_i|$. However, since the $\hat{e}_i$ are random variables themselves, some of them will be large even if there are no outliers, and a genuine outlier may not correspond to the largest $|\hat{e}_i|$, so we will occasionally declare the wrong case to be an outlier. This can most easily occur if v_{ii} for the true outlying point is large while the point declared to be an outlier has v_{ii} relatively small.

Remember that outlier identification is done relative to a specified linear model. If the form of the model is modified, the status of individual cases as outliers may change. Also, the detection of some outliers will have a greater effect on the regression estimates than will others. Cook's distance D_i provides relevant information on this point.

Testing for outliers. Suppose that the ith case is suspected to be an outlier. A useful procedure is done conceptually, but not computationally, as follows:

1. Delete the ith case from the data set.

2. Let $\hat{\boldsymbol{\beta}}_{-i}$ and $\hat{\sigma}^2_{-i}$ be the estimates of $\boldsymbol{\beta}$ and σ^2 using all the remaining $n - 1$ cases. $\hat{\sigma}^2_{-i}$ has $(n - 1) - p' = n - p' - 1$ degrees of freedom.

3. For the deleted case compute the value $\tilde{y}_i = \mathbf{x}_i^T \hat{\boldsymbol{\beta}}_{-i}$, using the symbol $\tilde{y}_i$ rather than $\hat{y}_i$ to suggest that we are using $\mathbf{x}_i^T \hat{\boldsymbol{\beta}}_{-i}$ to predict y_i since this case was not used in estimating $\boldsymbol{\beta}$. As the ith case was deleted from the data set before estimation, y_i and $\tilde{y}_i$ (as well as y_i and $\hat{\sigma}^2_{-i}$) are independent. The variance of $\tilde{y}_i$ is given by

$$\text{var}(\tilde{y}_i) = \sigma^2 \mathbf{x}_i^T (\mathbf{X}_{-i}^T \mathbf{X}_{-i}^T)^{-1} \mathbf{x}_i \tag{5.21}$$

where $\mathbf{X}_{-i}$ is the matrix $\mathbf{X}$ with the ith row deleted. $\text{Var}(\tilde{y}_i)$ is estimated by replacing σ^2 by $\hat{\sigma}^2_{-i}$ in (5.21).

4. Now, if y_i is not an outlier, $E(y_i - \tilde{y}_i) = 0$, but if y_i is an outlier, $E(y_i - \tilde{y}_i) \neq 0$. The quantity $y_i - \tilde{y}_i$ has variance $\text{var}(y_i - \tilde{y}_i) = \sigma^2(1 + \mathbf{x}_i^T (\mathbf{X}_{-i}^T \mathbf{X}_{-i})^{-1} \mathbf{x}_i)$ and, if we assume normal errors, a Student's t-test of the hypothesis $E(y_i - \tilde{y}_i) = 0$ is given by

$$t_i = \frac{y_i - \tilde{y}_i}{\hat{\sigma}_{-i}\sqrt{1 + \mathbf{x}_i^T (\mathbf{X}_{-i}^T \mathbf{X}_{-i})^{-1} \mathbf{x}_i}} \tag{5.22}$$

This test has $n - p' - 1$ degrees of freedom.

As was the case with the D_i's, it appears that a separate regression must be computed for each data point. However, as sketched in Appendix 5A.1, t_i can be computed as

$$t_i = r_i \left(\frac{n - p' - 1}{n - p' - r_i^2} \right)^{1/2} \tag{5.23}$$

and the t-test derived above is a monotonic transformation of the Studentized residual.

Some computer programs may print t_i instead of r_i or vice versa. If only the r_i's are available, tables of significance probabilities are given by Lund (1975). There is no important difference between using the t_i and the r_i in residual analysis.

Significance levels for the outlier test. If the investigator suspects in advance that the ith case is an outlier, then t_i should be compared to the central t-distribution with the appropriate number of degrees of freedom. Usually, the experimenter has no a priori choice for the outlier. If we test the case with the largest value of t_i to be an outlier, we are in reality performing n significance tests, one for each of n cases. Suppose, for example, that there were no outlier, and that $n = 65$, $p' = 5$. The probability that a t statistic with 60 degrees of freedom exceeds 2.000 (in absolute value) is .05; however, the probability that the largest of 65 independent t-tests exceeds 2.000 is .964, suggesting quite clearly the need for a different critical value. (Of course, the tests in our problem are correlated, so this computation is only a guide.*) The technique we use to find critical values is based on the *Bonferroni inequality*, which states that for n tests each of size α, the probability of falsely labeling at least one point an outlier is no greater than $n\alpha$. This procedure is conservative, since the Bonferroni inequality would specify only that the probability of the maximum of 65 tests exceeding 2.00 is no greater than 65(0.05), which is larger than 1. However, choosing the critical value to the $(\alpha/n) \times 100\%$ point of t will give a significance level of no more than $n(\alpha/n) = \alpha$. We would choose a level of $.05/65 = .00077$ for each test to give an overall level of no more than $65(.00077) = .05$.

To facilitate these computations of critical values, either special tables or charts of the extreme tails of the t-distribution or special tables of critical values, like Table D at the end of the book, are needed. To use Table D, select an α level (either .01 or .05 are available) and enter the table in the row corresponding to the sample size n and the column corresponding to the number of parameters p'. The corresponding entry in the table is $\alpha/n \times 100\%$ of t with $n - p' - 1$ degrees of freedom. For example, if $n = 39$, $p' = 5$, we would compare $|t_i|$ to 3.52 for a test at level $\alpha = 0.05$.

For example, in Forbes' data, Example 1.1, case number 12 was suspected to be an outlier because of its large residual. To perform the outlier test, we first need the Studentized residual, which is computed using (5.18) from $\hat{e}_i = 1.36$ (Table 1.5), $\hat{\sigma} = 0.379$ (Table 1.6), and $v_{12,12}$, which can be

---

*Excellent discussions of this and other multiple-test problems are presented by Miller (1965, 1977).

computed from (5.15) to be 0.0639. Thus

$$r_{12} = \frac{1.3592}{0.379\sqrt{1 - 0.0639}} = 3.7078$$

and the outlier test is

$$t_{12} = 3.7078\left(\frac{17 - 2 - 1}{17 - 2 - 3.7078^2}\right)^{1/2} = 12.40$$

This statistic has 14 degrees of freedom. Entering Table D with $p' = 2$ and $n = 17$, we find the critical value for a 0.01 test to be 4.41. Since t_i clearly exceeds this value, evidence is provided that this case is an outlier. However, as we have pointed out before, since $D_{12} = 0.47$ is not very large, this case has only a minor effect on estimating the parameters or other interesting aggregate computations. Notice, however, that inclusion of the outlier in the data does make the estimate of error variance much larger. Estimating this variance with suspected outliers included in the data would be a conservative procedure.

Problems

5.1 In a regression problem with $n = 54$, $p' = 5$, the results included $\hat{\sigma} = 2.0$, and the following case statistics for four of the cases.

$\hat{e}_i$	v_{ii}
0.6325	0.9000
1.732	0.7500
9.000	0.2500
10.295	0.0185

For each of these four cases, compute r_i, D_i, and t_i. Test each of the four cases to be an outlier. Make a qualitative statement about the influence of each case on the analysis.

5.2 In the fuel data as discussed in this chapter, test to see if Wyoming is an outlier (recall that the Studentized residual was 3.35, $n = 48$, $p' = 3$).

5.3 Using the QR factorization defined in Appendix 2A.3 and Problem 2.4, show that

$$V = Q_1 Q_1^T = I - Q_2 Q_2^T$$

$$I - V = Q_2 Q_2^T = I - Q_1 Q_1^T$$

Hence, if q_i^T is the ith row of Q_1,

$$v_{ij} = q_i^T q_j \quad \text{and} \quad v_{ii} = q_i^T q_i$$

Thus if the QR factorization of $\mathbf{X}$ is computed, the v_{ij} and the v_{ii} are easily computed as well.

5.4   Let $\mathbf{U}$ be an $n \times 1$ vector with a 1 as its first element, and zeros elsewhere. Consider computing the regression of $\mathbf{U}$ on $\mathbf{X}$, where $\mathbf{X}$ is an $n \times p'$ rank p' data matrix. Let $\mathbf{V} = \mathbf{X}(\mathbf{X}^T\mathbf{X})^{-1}\mathbf{X}^T$ with elements v_{ij}.

  5.4.1. Show that the vector of fitted values from the regression of $\mathbf{U}$ on $\mathbf{X}$ are the $v_{1j}, j = 1, \ldots, n$.

  5.4.2. Show that the vector of residuals are $1 - v_{11}$ and $-v_{1j}, j > 1$.

  5.4.3. What are the Studentized residuals?

Using this method, the v_{ij}'s can be found in any regression program that computes residuals.

5.5   Two $n \times n$ matrices $\mathbf{A}$ and $\mathbf{B}$ are orthogonal if $\mathbf{AB} = \mathbf{BA} = 0$. Show that $\mathbf{I} - \mathbf{V}$ and $\mathbf{V}$ are orthogonal. Use this result to show that as long as the intercept is in a regression model the sample correlation between the residuals $\hat{\mathbf{e}} = (\mathbf{I} - \mathbf{V})\mathbf{Y}$ and the fitted values $\hat{\mathbf{Y}} = \mathbf{VY}$ is exactly zero, or equivalently that the slope of the regression of $\hat{\mathbf{e}}$ on $\hat{\mathbf{Y}}$ is zero.

5.6   The matrix $\mathbf{X}_{-i}^T\mathbf{X}_{-i}$ can be written as $\mathbf{X}_{-i}^T\mathbf{X}_{-i} = \mathbf{X}^T\mathbf{X} - \mathbf{x}_i\mathbf{x}_i^T$ where $\mathbf{x}_i$ is the ith row of $\mathbf{X}$. Use this definition to prove that (5A.1) holds.

5.7   The quantity $y_i - \mathbf{x}_i^T\hat{\boldsymbol{\beta}}_{-i}$ is the residual for the ith case when $\boldsymbol{\beta}$ is estimated without the ith case. Use (5A.1) to show that

$$y_i - \mathbf{x}_i^T\hat{\boldsymbol{\beta}}_{-i} = \frac{\hat{e}_i}{1 - v_{ii}}$$

This quantity is called the predicted residual or PRESS, and has been used in various procedures in regression (Allen, 1974).

5.8   Use (5A.1) to verify (5.20).

5.9   Suppose that interest centered on $\boldsymbol{\beta}^*$ rather than $\boldsymbol{\beta}$ (where as in Chapters 2 and 4, $\boldsymbol{\beta}^*$ is the parameter vector excluding β_0). Using (4.26) as a basis, define a distance measure D_i^* analogous to Cook's D_i, and show that (Cook, 1979)

$$D_i^* = \frac{r_i^2}{ps^2}\left(\frac{v_{ii} - 1/n}{1 - v_{ii}}\right)$$

5.10   Use (5A.1) to prove (5.23), the relationship between r_i and t_i. What is the largest possible value of r_i?

6

CASE ANALYSIS II:
SYMPTOMS AND REMEDIES

Examination of lists of the statistics defined in Chapter 5—Studentized residuals r_i, v_{ii}, Cook's D_i, and the outlier test t_i—can provide information about the adequacy of a fitted model. In this chapter, we study the ways in which a model can fail and examine diagnostic procedures to find such failures as heterogeneity of error variance, nonlinearity of the model, non-normality of errors, outliers, or inadequacy of the hypothesized model. Remedies suited to the symptoms are then suggested. In many contexts in data analysis, one should view this methodology as one would view a physician treating a patient for a disease on the basis of observed symptoms. Often, identical symptoms are indicative of more than one problem, and a given remedy may be appropriate for the symptoms, but not for the cause. On the other hand, the remedy may produce side effects that are little better than the original problem. Unlike the physician, however, the statistician may apply a remedy, assess the results, usually by repeating the analysis, and, if the remedy is not an improvement, reconsider its use. In many applications in data analysis, it is both reasonable and proper to try sequences of potential remedies—several transformations, deletion of variables or cases, and so on—especially if the goal of the analysis is to gain an understanding of the ways in which a set of variables interact. However, this may not be appropriate if the objective of the analysis is the testing of specific hypotheses, since the characteristics of test statistics after the techniques of data analysis are applied are essentially unknown.

Some model inadequacies will be more serious than others. For example, non-normality of errors may serve only to make tabled percentage points for the test statistics invalid. Other types of failures are potentially more

119

serious. Heterogeneity of error variances, for example, can make the estimation procedure less than optimal unless it is diagnosed and procedures are adjusted accordingly. More serious still is nonlinearity of the model. If we fit a nonlinear model by assuming that it is linear, applicability of the model to future cases and interpretation of the usual aggregate statistics is questionable. An incorrectly specified model will lack repeatability, and the linear model we fit will give aggregate statistics appropriate for the observed data, but another experimenter using different cases may obtain different aggregate statistics, possibly leading to contradictory conclusions.

Most of the methods that are useful in finding failures in a model are based on the statistics developed in Chapter 5. Often the diagnostics are graphical, so it is essential to study a variety of plots or graphs of the original data (the independent variables $X_1, \ldots, X_p$ and the dependent variable Y), derived statistics (especially the Studentized residuals r_i and the fitted values $\hat{y}_i$), and, occasionally, other variables such as time or case number.

The importance of producing and analyzing plots as a standard part of statistical analysis cannot be overemphasized. Besides occasionally providing an easy to understand summary of a complex problem, they allow the simultaneous examination of the data as an aggregate while clearly displaying the behavior of individual cases.

6.1   Diagnostics: plotting residuals

Plots of residuals are the most important diagnostic in case analysis. The properties of residuals for a correctly specified model are given in Chapter 5. When the statistical model fails, some of the properties of the residuals will not hold, and these failures may be reflected in residual plots.

In the residual plots considered here, the Studentized residuals will always be plotted against other variables or functions of the data since there appears to be little that is lost, and much that is gained, by using them rather than the ordinary residuals. Recall that if the model used to compute the r_i is valid, each r_i has mean 0 and variance 1. If the form of a model is correct, and we assume normality of errors, then about 95% of the r_i's should fall in the range $+2$ to -2, while only about one r_i in 1000 should fall outside the range $+3$ to -3. When the r_i's are plotted against a function of the data such as $\hat{y}_i$, the plot should exhibit no systematic trends. If such trends are observed, or large $|r_i|$ are observed, the model or some part of it may be suspect.

Plots against $\hat{y}_i$. The most important single diagnostic is the plot of r_i versus $\hat{y}_i$. From Section 5.1, the r_i's and $\hat{y}_i$'s are nearly uncorrelated, so the

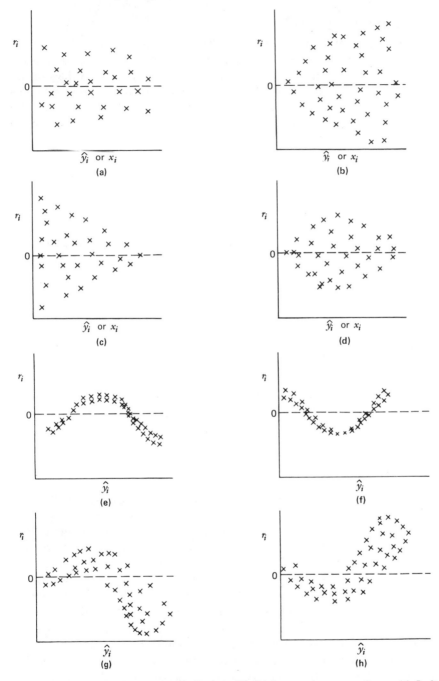

Figure 6.1 Residual plots. **(a)** Null plot. **(b)** Right-opening megaphone. **(c)** Left-opening megaphone. **(d)** Double outward bow. **(e)** Nonlinearity. **(f)** Nonlinearity. **(g)** Nonlinearity and nonconstant variance. **(h)** Nonlinearity and nonconstant variance.

null plot, the plot that is observed when the specified model is correct, might look something like Figure 6.1*a*. In residual plotting, the user is primarily interested in the shape of the plot, and often the values on the axes are not important. Since the vertical axis is for Studentized residuals, the range $(+3, -3)$ is generally adequate. In Figure 6.1*a*, for example, the points tend to fall in a horizontal band, without any apparent systematic features.

Plots against independent variables. Like the fitted values, the r_i's and each of the independent variables are nearly uncorrelated, and systematic features in these plots would suggest model failures that are a function of the independent variable plotted. For example, observing the right-opening megaphone, Figure 6.1*b*, in a plot of the r_i against an independent variable may suggest that the residual variance is an increasing function of that independent variable.

Other plots. In some problems, plotting r_i versus other quantities such as case number or time may be of use, as these could indicate failures of a model that are due to ordering or changes over time. These plots are of a different character than plots against $\hat{y}_i$ or the independent variables, as the r_i's may be highly correlated with these extra variables, and very systematic plots may be observed. Sometimes, precisely these systematic features in the plots are of interest.

6.2   Heterogeneity of variance

In many practical and theoretical contexts, the assumption that the error variance is constant for each data point (e.g., $\text{var}(e_i) = \sigma^2$) is uncertain. Commonly, the magnitude of $\text{var}(e_i)$ will depend upon the magnitude of some other variable, often the response. For example, if an intrinsically positive response varies over a wide range, say from near zero into the thousands, it is intuitively clear that responses near zero usually will be less variable than responses near 1000, since the latter have more "room" to vary than do the former. This can arise in experimental situations in which the control units, without any treatments, have very small response, but addition of a treatment results in a large, but variable, response, or in a drug trial where poorly treated units die immediately, but well treated units live a long, but variable, time.

Symptoms. Residual plots like those of Figure 6.1*b* through 6.1*d* indicate that the error variance is a systematic function of the quantity plotted on the horizontal axis. In the situation described above, the right-opening megaphone pattern of Figure 6.1*b* should be expected when the horizontal axis is $\hat{y}_i$. Also, if var(e_i) is strongly related to any good predictor of Y, say X, then the plot of r_i against X should reveal this as well.

The left-opening megaphone (Figure 6.1*c*) will occur when small values of the horizontal axis imply large variability. The double outward bow, Figure 6.1*d*, can arise if the response is a percentage between 0 and 100%. Large or small percentages are less variable than are percentages near 50%.

Generally, residual plots will not be as clear as those in Figure 6.1. The eye will often be influenced by a few points, and heterogeneity of variance may be suspected when the problem is actually an outlier. Also, cases with extreme values of the variable plotted on the x-axis are often more carefully examined as indicators of heterogeneity. Yet these are generally the points that have the greatest influence in the estimation of parameters (Section 5.2) and will therefore tend to be overfit, that is, have residuals that are relatively small. This problem is partially solved by using the r_i's rather than the $\hat{e}_i$'s, but the problem remains if the fitted model is incorrect.

Remedies. If heterogeneity of variance is suspected, and var(e_i) is a known function of some quantity, such as var(e_i) = $\sigma^2 z_i$, where z_i is known for each case (z_i may be an independent variable, or some other variable such as time), then weighted least squares is suggested (Section 4.1). As a result, best linear unbiased estimates of the parameter vector can be obtained. Often, however, the heterogeneity of variance can be removed by transforming either the response Y or one or more of the X's or both; usually, transformations to Y will be applied. These transformations are called *variance stabilizing transformations*.

For almost any relationship between var(e_i) and a response, an appropriate variance stabilizing transformation can be found (see Scheffe (1959), Chapter 9 for technical details). In Table 6.1, the common variance stabilizing transformations are listed. The first three ($Y^{1/2}$, log(Y), $1/Y$), as well as their forms when some of the responses are zero, are appropriate for the right- (or perhaps left-) opening megaphone form, but each is more severe than the one before it. The square-root transformation is relatively mild and is most appropriate when the y_i's are counts following a Poisson distribution, usually the first model considered for errors in counts. The logarithm is the most commonly used transformation (the base is irrelevant). It is exactly appropriate when the error standard deviation is a percentage or a proportion of the response (i.e., the error is $\pm 10\%$, not ± 10 units), and $[\text{var}(e_i)]^{1/2} \propto E(y_i)$. More will be said about such models in the next section.

Table 6.1   Common variance stabilizers

Transformation	Situation	Comments
$\sqrt{Y}$	$\text{var}(e_i) \propto E(Y_i)$	The theoretical basis is for counts from the Poisson distribution
$\sqrt{Y} + \sqrt{Y+1}$	As above	For use when some Y_i's are zero or very small; this is called the Free-man-Tukey (1950) transformation
$\log Y$	$\text{var}(e_i) \propto [E(Y_i)]^2$	This transformation is very common; it is a good candidate if the range of Y is very broad, say from 1 to several thousand; all Y_i must be strictly positive
$\log(Y+1)$	As above	Used if $Y_i = 0$ for some cases
$1/Y$	$\text{var}(e_i) \propto [E(Y_i)]^4$	Appropriate when responses are "bunched" near zero, but, in markedly decreasing numbers, large responses do occur; e.g., if the response is a latency or response time for a treatment or a drug, some subjects may respond quickly while a few take much longer; the reciprocal transformation changes the scale of time per response to the rate of response, response per unit time; all Y_i must be positive
$1/(Y+1)$	As above	Used if $Y_i = 0$ for some cases
$\sin^{-1}(\sqrt{Y})$	$\text{var}(e_i) \propto E(Y_i)(1 - E(Y_i))$	For binomial proportions $(0 \leqslant Y_i \leqslant 1)$

The reciprocal (or inverse) transformation is often applied when the response is a time of waiting, healing, survival, and so on. Taking reciprocals changes the scale from time per response to responses per unit time. This latter is a rate and it is often more interesting on theoretical grounds than the original measurement.

When repeated measurements are made at each of several values of **x**, additional information concerning nonconstant variance is available, since an estimate of a variance at each **x** can then be computed. For example, consider again the apple shoot data in Example 4.2. There, for each of the sample days, the mean and standard deviation of the number of stem units observed were recorded. They are plotted in Figure 6.2. From the plot, it appears that the standard deviation is an increasing function of the mean, suggesting the need for transformation of Y. From Table 6.1, the choices of $\log(Y)$ or $Y^{1/2}$ appear to be good candidates. Unfortunately, the original data are not available, so we cannot actually check the transformations and see how well they work.

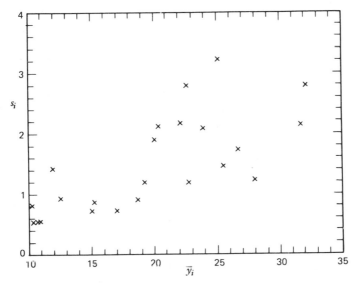

Figure 6.2 Day averages versus standard deviation for apple shoot data.

Table 6.2 Linearizing transformations

Transformation		Simple Regression Form	Multiple Regression Form
$\log Y$	$\log X$	$Y = \alpha X^{\beta}$	$Y = \alpha X_1^{\beta_1} X_2^{\beta_2} \cdots X_p^{\beta_p}$
$\log Y$	X	$Y = \alpha e^{\beta X}$	$Y = \alpha e^{\sum \beta_j X_j}$
Y	$\log X$	$Y = \alpha + \beta(\log X)$	$Y = \alpha + \sum \beta_j \log(X_j)$
$\dfrac{1}{Y}$	$\dfrac{1}{X}$	$Y = \dfrac{X}{\alpha X + \beta}$	$Y = \dfrac{1}{\alpha + \sum(\beta_j / X_j)}$
$\dfrac{1}{Y}$	X	$Y = \dfrac{1}{\alpha + \beta X}$	$Y = \dfrac{1}{\alpha + \sum \beta_j x_j}$
Y	$\dfrac{1}{X}$	$Y = \alpha + \beta\left(\dfrac{1}{X}\right)$	$Y = \alpha + \sum \beta_j\left(\dfrac{1}{X_j}\right)$

6.3   Nonlinearity

Not all relationships between a response and a set of predictors are linear; in fact, we might expect linear relationships to be the exception rather than the rule. However, the usefulness of a linear model is more general than is apparent. For example, while a functional form may be nonlinear over the entire range of the independent variables, it may behave as if it were approximately linear over restricted ranges for them; different restricted ranges may be adequately described by different approximating linear models. If we fit a linear model that is appropriate for a specific range for the independent variables, inference to other values of the independent variables is rarely reliable, and often quite misleading. However, useful information inside the relevant region can often be obtained and used in a meaningful way.

More importantly, suitable transformations of data can frequently be found that will reduce a theoretically nonlinear model to a linear form. These transformations are said to be linearizable and comprise a class of functions that may either occur in practice or may themselves provide reasonable approximations to functions that occur in practice. Linearizing may require transforming both the independent and the dependent variable.

An important class of linearizable functions are the power or multiplicative models of the form (for one independent variable)

$$Y = \alpha X^\beta \tag{6.1}$$

Some members of this family for fixed $\alpha = 1$ and varying β are shown in Figure 6.3. This form is linearized by taking logarithms,

$$\log Y = \log \alpha + \beta \log X \tag{6.2}$$

so that the regression of $\log Y$ on $\log X$ is linear. Some attention, however, should be paid to errors. Multiplicative errors of the form "k percent of the response" can be incorporated into (6.2) by writing

$$Y = (\alpha X^\beta)(e) \tag{6.3}$$

so that the observed response is a percentage of the true response multiplied by an error. Taking logarithms,

$$\log Y = \log \alpha + \beta \log X + \log e \tag{6.4}$$

giving an additive error in the log scale. This model is in the form of the usual linear model, and can be studied by usual techniques.

Table 6.2 lists some of the common linearizable forms and the appropriate transformations needed to achieve linearity. If any of these forms are reasonable on theoretical grounds, the given transformations should be used. However, in many situations in regression, the transformations of data to be used cannot be chosen theoretically but must be determined as part of the data analysis.

Not all functions are linearizable, nor in some cases is it desirable to transform for linearity. As an example of the former, a sum of two exponentials,

$$Y = \alpha_1 e^{-\beta_1 X_1} + \alpha_2 e^{-\beta_2 X_2}$$

is not linearizable. As an example of the latter, the logistic function,

$$Y = \frac{e^{-(\alpha + \beta X)}}{1 + e^{-(\alpha + \beta X)}}$$

for $0 \leqslant Y \leqslant 1$ is linearizable by the logit transformation $\ln[Y/(1 - Y)]$, but applying linear model methods to this transformed problem has poor statistical properties. Both of these models are best handled via nonlinear methods; for discussion of the logistic model, see Cox (1970) or Feinberg (1977).

Symptoms of nonlinearity. Nonlinearity can often be discovered by examining plots of the r_i versus the fitted values $\hat{y}_i$ or other variables for systematic trends, perhaps like those in Figures 6.2e through 6.2h. In general, nonlinearity will be indicated by a curved trend in the r_i's when plotted against $\hat{Y}$ or one of the X's; the curve may open upward or

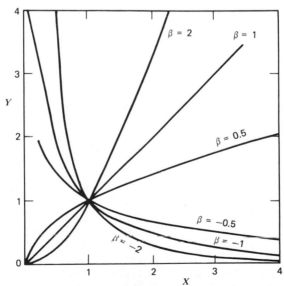

Figure 6.3 A family of power curves, $Y = X^\beta$.

downward, depending on the values of unknown parameters. Thus in practical work, the specific choice of a transformation to achieve linearity will depend largely on the nature of the variables, and other consider- ations.

If repeated measurements are made at each value of X, the F-test for lack of fit described in Section 4.3 is appropriate, and provides a strong symptom for nonlinearity. However, the F-statistic must be viewed as a guide, not as a test statistic, as the exploratory procedure may invalidate the distributional assumptions of the F-test.

Remedies. The principle remedies for nonlinearity are transformations, and addition of polynomial and, occasionally, cross-product terms in the X's already in the model. The symptoms that call for polynomials or for linearizing transformations are often identical. Guidance from knowledge of other sources may be needed, such as from the error structure or nature of the measurements.

Example 6.1   Brain and body weights

Table 6.3 gives the average brain and body weights for 62 species of mammals. We shall consider the problem of modeling brain weight as a function of body weight. These data were taken from a larger study, and were collected for another purpose (Allison and Cicchetti, 1976).

Table 6.3   Average brain weights and body weights of 62 species of mammals

Species (Common Name)	Body weight (kg)	Brain weight (g)	log(body weight)	log(brain weight)
1 Arctic fox	3.385	44.500	0.530	1.648
2 Owl monkey	0.480	15.500	−0.319	1.190
3 Mountain beaver	1.350	8.100	0.130	0.908
4 Cow	465.000	423.000	2.667	2.626
5 Gray wolf	36.330	119.500	1.560	2.077
6 Goat	27.660	115.000	1.442	2.061
7 Roe deer	14.830	98.200	1.171	1.992
8 Guinea pig	1.040	5.500	0.017	0.740
9 Vervet	4.190	58.000	0.622	1.763
10 Chinchilla	0.425	6.400	−0.372	0.806
11 Ground squirrel	0.101	4.000	−0.996	0.602
12 Arctic ground squirrel	0.920	5.700	−0.036	0.756
13 African giant pouched rat	1.000	6.600	−0.000	0.820
14 Lesser short-tailed shrew	0.005	0.140	−2.301	−0.854
15 Star nosed mole	0.060	1.000	−1.222	−0.000
16 Nine-banded armadillo	3.500	10.800	0.544	1.033
17 Tree hyrax	2.000	12.300	0.301	1.090

Table 6.3 (continued)

	Species (Common Name)	Body weight (kg)	Brain weight (g)	log(body weight)	log(brain weight)
18	N. American opossum	1.700	6.300	0.230	0.799
19	Asian elephant	2547.000	4603.000	3.406	3.663
20	Big brown bat	0.023	0.300	− 1.638	− 0.523
21	Donkey	187.100	419.000	2.272	2.622
22	Horse	521.000	655.000	2.717	2.816
23	European hedgehog	0.785	3.500	− 0.105	0.544
24	Patas monkey	10.000	115.000	1.000	2.061
25	Cat	3.300	25.600	0.519	1.408
26	Galago	0.200	5.000	− 0.699	0.699
27	Genet	1.410	17.500	0.149	1.243
28	Giraffe	529.000	680.000	2.723	2.833
29	Gorilla	207.000	406.000	2.316	2.609
30	Gray seal	85.000	325.000	1.929	2.512
31	Rock hyrax[a]	0.750	12.300	− 0.125	1.090
32	Man	62.000	1320.000	1.792	3.121
33	African elephant	6654.000	5712.000	3.823	3.757
34	Water opossum	3.500	3.900	0.544	0.591
35	Rhesus monkey	6.800	179.000	0.833	2.253
36	Kangaroo	35.000	56.000	1.544	1.748
37	Yellow-bellied marmot	4.050	17.000	0.607	1.230
38	Golder hamster	0.120	1.000	− 0.921	− 0.000
39	Mouse	0.023	0.400	− 1.638	− 0.398
40	Little brown bat	0.010	0.250	− 2.000	− 0.602
41	Slow loris	1.400	12.500	0.146	1.097
42	Okapi	250.000	490.000	2.398	2.690
43	Rabbit	2.500	12.100	0.398	1.083
44	Sheep	55.500	175.000	1.744	2.243
45	Jaguar	100.000	157.000	2.000	2.196
46	Chimpanzee	52.160	440.000	1.717	2.643
47	Baboon	10.550	179.500	1.023	2.254
48	Desert hedgehog	0.550	2.400	− 0.260	0.380
49	Giant armadillo	60.000	81.000	1.778	1.908
50	Rock hyrax[b]	3.600	21.000	0.556	1.322
51	Raccoon	4.288	39.200	0.632	1.593
52	Rat	0.280	1.900	− 0.553	0.279
53	Eastern American mole	0.075	1.200	− 1.125	0.079
54	Mole rat	0.122	3.000	− 0.914	0.477
55	Musk shrew	0.048	0.330	− 1.319	− 0.481
56	Pig	192.000	180.000	2.283	2.255
57	Echidna	3.000	25.000	0.477	1.398
58	Brazilian tapir	160.000	169.000	2.204	2.228
59	Tenrec	0.900	2.600	− 0.046	0.415
60	Phalanger	1.620	11.400	0.210	1.057
61	Tree shrew	0.104	2.500	− 0.983	0.398
62	Red fox	4.235	50.400	0.627	1.702

[a] *Heterohyrax brucci.*
[b] *Procavia habessinica.*

The initial attempt to graph brain weight (in grams) versus body weight (in kilograms), as given in Figure 6.4, indicates immediately that some sort of transformation is required. Most of the points in the plot are jammed into the lower left-hand corner with only a few stragglers elsewhere. Because of the wide variation of both variables, log transformations are obvious candidates. This is like assuming that the correct functional relationship is of the form brain weight = α_0(body weight)$^{\alpha_1}$. The logarithms (to the base 10) of the weights are also given in Table 6.3. The scatter plot in the log scale is given in Figure 6.5, suggesting that there is a strong linear relationship in the log scale. The scatter plot of the residuals from the fitted regression is given in Figure 6.6. The fitted line corresponds to the observed data remarkably well. The regression summaries are given in Table 6.4.

In this example, the transformation used seems to achieve two important goals, namely, linearity and constant variance. In many problems, the transformation appropriate for one of these will not be the appropriate transformation for the other. Note, also, from Figure 6.6 that the largest residual is for Man, with $r_i = 2.848$ and t_i (from (5.23)) equal to 3.04. To attach a significance level to this statistic, the Bonferroni t-tables could be used; however, the interest in Man as a special case is a priori, so the usual t-tables can be used. The appropriate significance level is then, from Table A, about 0.005, so Man would be declared an outlier, and this species has a larger than predicted average brain weight.

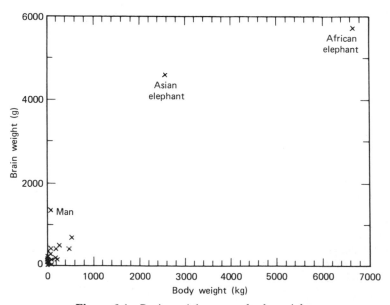

Figure 6.4 Brain weight versus body weight.

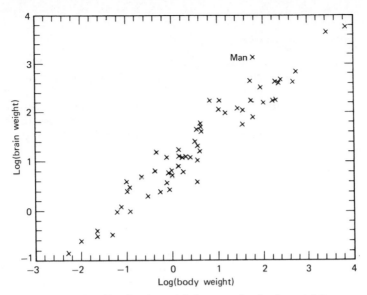

Figure 6.5 Log(brain weight) versus log(body weight).

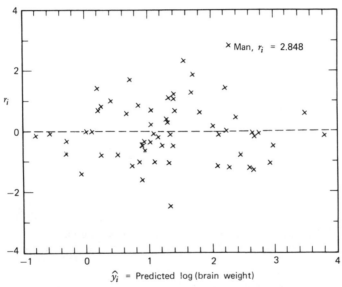

Figure 6.6 Residual plot, brain weight–body weight data.

Table 6.4 Regression summary: brain weight/body weight data

	Estimate	Standard Error	t-Value
Intercept	0.927	0.0417	22.23
Slope	0.752	0.0285	26.41

$R^2 = 0.92$, $\hat{\sigma} = 0.0191$, d.f. $= 60$

It is important to note that the model fit is for species averages and may not apply to individuals in a species. However, if data were available on individuals in a species, then the model used here could be fit separately for each. The fit of these models is generally excellent, and some ecologists use the parameter estimates for this model to summarize the relationship between brain weight and body weight within a species.

6.4   Distributional assumptions

The usual distributional assumption in regression analysis is that the errors are normally distributed. Normal distribution of errors is needed to justify F and t testing and confidence interval procedures. The problem of non-normality of errors is very difficult to diagnose by examination of residuals. To see this, suppose we have a linear model $\mathbf{Y} = \mathbf{X}\boldsymbol{\beta} + \mathbf{e}$, where $\text{cov}(\mathbf{e}) = \sigma^2\mathbf{I}$, and we wish to examine the assumption of the normality of the errors. Of course, $\mathbf{e}$ is never observed, so the test of normality must be based on the residuals

$$\hat{\mathbf{e}} = \mathbf{Y} - \hat{\mathbf{Y}} = \mathbf{Y} - \mathbf{X}(\mathbf{X}^T\mathbf{X})^{-1}\mathbf{X}^T\mathbf{Y}$$

Substituting $\mathbf{X}\boldsymbol{\beta} + \mathbf{e}$ for $\mathbf{Y}$ in this last expression,

$$\hat{\mathbf{e}} = \mathbf{X}\boldsymbol{\beta} + \mathbf{e} - \mathbf{X}(\mathbf{X}^T\mathbf{X})^{-1}\mathbf{X}^T(\mathbf{X}\boldsymbol{\beta} + \mathbf{e})$$

$$= \mathbf{X}\boldsymbol{\beta} + \mathbf{e} - \mathbf{X}\boldsymbol{\beta} - \mathbf{X}(\mathbf{X}^T\mathbf{X})^{-1}\mathbf{X}^T\mathbf{e}$$

$$= \left[\mathbf{I} - \mathbf{X}(\mathbf{X}^T\mathbf{X})^{-1}\mathbf{X}^T\right]\mathbf{e} \tag{6.5}$$

and the test for normality of $\mathbf{e}$ will be based on $\hat{\mathbf{e}}$. Recalling that $\mathbf{V} = \mathbf{X}(\mathbf{X}^T\mathbf{X})^{-1}\mathbf{X}^T$, and v_{ij} is the (i, j)th element of $\mathbf{V}$, the last equation can be written

$$\hat{\mathbf{e}} = (\mathbf{I} - \mathbf{V})\mathbf{e} \tag{6.6}$$

or, in scalar form, $\hat{e}_i$ is

$$\hat{e}_i = e_i - \left(\sum_{j=1}^{n} v_{ij}e_j\right) \tag{6.7}$$

Thus $\hat{e}_i$ is equal to e_i adjusted by subtracting off a weighted sum of all the e_j's (including e_i) with $v_{ij} \neq 0$. If the number of degrees of freedom for error $(n - p')$ is small, and some of the v_{ij}'s are large, as will be the case in many regression problems, the term in parentheses in (6.7) may be more important than the term e_i, and hence $\hat{e}_i$ will behave more like $\sum v_{ij}e_j$ than like

e_i; this sum will tend to behave like a normal random variable even if the e_j are not normal. Thus any test for non-normality applied to the $\hat{e}_i$ (or to the r_i) cannot be expected to be very good. Gnanadesikan (1977) refers to this as the supernormality of residuals.

As n increases (for fixed p'), the v_{ij} will tend towards zero and hence the term e_i will dominate in (6.7); the second term $\sum v_{ij}e_j$ will still tend to a normal variate, but its asymptotic variance will be $\sigma^2 \sum v_{ij}^2 = \sigma^2 v_{ii}$, which will tend to be small relative to the assumed $\text{var}(e_i) = \sigma^2$. Thus if $n - p'$ is sufficiently large, "usual" techniques for assessing normality may be applied.

Probability plots. Our technique of choice for studying non-normality is the normal probability or *rankit* plot (a general treatment of probability plotting is given by Wilk and Gnanadesikan (1968) and Gnanadesikan (1977)). Suppose we have a sample of n numbers $z_1, z_2, \ldots, z_n$, and we wish to examine the chance that the z's are a homogeneous sample from a normal distribution with unknown mean μ and variance σ^2. A useful way to proceed is as follows:

1. Order the z's to get $z_{(1)} \leqslant z_{(2)} \leqslant \cdots \leqslant z_{(n)}$. The ordered z's are called the sample order statistics.

2. Now, consider a normal sample of size n with zero mean and unit variance. Let $u_{(1)} \leqslant u_{(2)} \leqslant \cdots \leqslant u_{(n)}$ be the mean values of the order statistics that would be obtained if we repeatedly took samples of size n from the standard normal. The $u_{(i)}$'s are called the expected values of normal order statistics, or *rankits*. The rankits are frequently tabled, as in Table E, or can be easily approximated using a computer.*

3. If the z's are normal, then

$$E(z_{(i)}) = \mu + \sigma u_{(i)}$$

so that the regression of $z_{(i)}$ on $u_{(i)}$ will be a straight line. If the sample is not normal, then the rankit plot should not approximate a straight line.

The plot of $z_{(i)}$ versus $u_{(i)}$ is called a rankit plot. Rankit plots can also be made using *normal probability paper*. Then, the $z_{(i)}$'s are graphed versus the values $i/(n + 1)$, and the probability scale on the paper will then result in approximations to the $u_{(i)}$'s.

---

* Using a subroutine that computes the cumulative normal distribution function $\Phi(t) = \int_{-\infty}^{t} (2\pi)^{-1/2} \exp(-x^2/2)\, dx$ and its inverse $\Phi^{-1}(p)$, a simple approximation to $u_{(i)}$ for a sample of size n is $u_{(i)} \doteq \Phi^{-1}(i - \frac{3}{8})/(n + \frac{1}{4})$; see Blom (1958) or Weisberg and Bingham (1975).

As an example of a rankit plot, a pseudorandom sample was generated on a computer with $n = 17$ as if the sample were drawn from $N(3, 4)$. Using the rankits from Table E, the rankit plot in Figure 6.7 was drawn. This plot is nearly, though not exactly, a straight line. In contrast, Figure 6.8 is a rankit plot of $n = 17$ pseudorandom numbers drawn as if they were from a uniform distribution on the interval $(0, 1)$. This latter plot exhibits a clear flattening at both ends of the plot—suggesting that too many values relatively far from the mean are in the sample for it to be considered a sample from a normal distribution. Besides the $\int$ observed in this plot, other common shapes for non-normal distributions include $\frown$, $\int$, and $\underline{\diagup}$. The first of these indicates too few extreme values, while the remaining two shapes indicate negative and positive skew, respectively.

In general, judging whether or not a rankit plot indicates that a sample behaves as if it were normal requires an experienced observer. Daniel and Wood (1971) and Daniel (1976) provide many pages of "training plots" that may help the analyst learn to interpret rankit plots.

The remedy for non-normality is transformation of the dependent variable. Unfortunately, the transformation appropriate to achieve normality will often sacrifice linearity or constant variance. Of the three, normality is least important in the analysis.

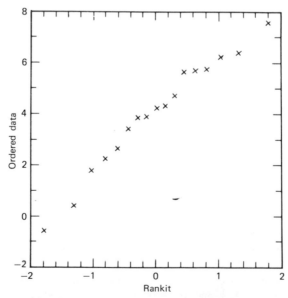

Figure 6.7 Rankit plot with normal data.

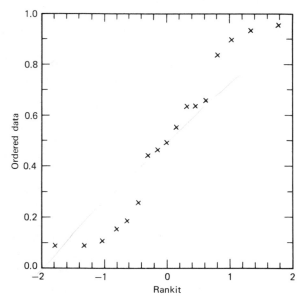

Figure 6.8   Rankit plot with uniform data.

Correlated errors. Another distributional assumption concerning the errors is that they are uncorrelated. This would mean that the value of the error for one case does not depend on the value of the error for any other case. This assumption will be violated in some problems, especially if the cases are ordered in time or in space, and adjacent cases influence each other.

Diagnostics for finding correlated errors are usually applicable only in special circumstances. For example, if cases are equally spaced in time, then the Durbin-Watson statistic (Durbin and Watson, 1950, 1951, 1971) is appropriate for testing for correlation between adjacent cases. In general, diagnosing correlated errors is very difficult, the best diagnostics coming from careful consideration of the process that generates the data.

6.5   Outliers and extreme cases

As in Section 5.3, outliers are cases for which the hypothesized model fails to fit, while the model is satisfactory for all or most of the other cases. The candidates for outliers are cases with extreme values of the Studentized residual r_i; an appropriate test statistic was discussed in Section 5.3. One should check the r_i for large values as a standard practice. The importance of detecting an outlier, and actually doing something about it (e.g., deleting the case and repeating the analysis) depends both on the magnitude of r_i and on the magnitude of v_{ii} or more simply on the size of Cook's distance D_i. If D_i is quite small, then the outlier (if it exists) has little effect on the estimates of the parameters, and it is therefore not important to exclude the case from the analysis even if it is declared to be an outlier on the basis of a test. On the other hand, cases with large values of v_{ii} may deserve special attention even if the corresponding r_i's are small since if D_i is large, the case may have an undue influence on the estimates of the parameters. We may thereby get a false picture of the regression surface.

The remedies for data sets with suspected outliers or influential cases are more complex than the other remedies proposed in this chapter. A generally useful step is to refit the regression without the suspected case(s) and assess the changes in the analysis (this would usually include repeating the residual analysis). This is a reasonable approach, especially if the form of the model (e.g., the scale of all the X's) is known to be correct. But, if the correct scale for the variables is not known, then cases with large values of v_{ii}, D_i, or r_i may be indicative of the need for a transformation, not an outlier. For example, suppose that one of the variables is family income. If a regression analysis is performed, and most cases have incomes under $30,000, but one case has an income of about $1,000,000, that case may be very influential in the estimation procedure. However, if income is reexpressed as log(income), where $\log(30{,}000) = 4.48$ and $\log(1{,}000{,}000) = 6.0$, then, since in the log scale the incomes are not that different, the higher income case would not be as influential (in the sense of influencing the estimates of parameters!).

Thus the detection of potential outliers or influential cases may properly result in either of two remedies. First, we may refit the regression without the offending case and see exactly what happens. Secondly, we may consider reasonable transformations of the data that would appear to make influential cases less important or improve the fit of the regression to the cases that are suspected outliers.

6.6 Choosing a transformation

The methods for choosing transformations discussed so far require either specific knowledge about the relationships between variables, or use diagnostics to suggest possible transformations. There are more objective methods for determining appropriate transformations. These generally will choose a transformation to maximize some criterion function of interest. Generally, use of these methods will require additional assumptions concerning the data, often normality of errors. Two methods are described here. The first, proposed by Box and Cox (1964), is appropriate for transforming the response Y, while the second, generally modeled after proposals by Box and Tidwell (1962), is appropriate for transforming the X's.

Transforming Y: power families. An important class of transformations are those obtained, when Y can only take positive values, by raising the observed data to a power. If we consider only transformations of Y, this is equivalent to fitting the model

$$\mathbf{Y}^\lambda = \mathbf{X}\boldsymbol{\beta} + \mathbf{e}, \quad \text{var}(\mathbf{e}) = \sigma^2 \mathbf{I} \tag{6.8}$$

where λ is the power of the transformation ($\mathbf{Y}^\lambda$ means take each element of $\mathbf{Y}$ and raise it to the λ power; i.e., $\mathbf{Y}^2$ means square each element of $\mathbf{Y}$). If, by convention, we take $\lambda = 0$ to be the (natural) log transformation, then all the usual transformations are included: $\lambda = 1$ corresponds to no transformation, $\lambda = -1$ corresponds to reciprocals, $\lambda = \frac{1}{2}$ to square root.

When the appropriate λ is unknown, we can view (6.8) as specifying a model that has one additional unknown parameter, λ, in addition to unknown $\boldsymbol{\beta}$ and σ^2. Thus, we can estimate λ just as we can estimate $\boldsymbol{\beta}$ and σ^2, from observed data. Box and Cox (1964) have suggested that an estimate of λ can be obtained by finding the joint *maximum likelihood estimate* $(\hat{\boldsymbol{\beta}}, \hat{\sigma}^2, \hat{\lambda})$ using distributional assumptions concerning $\mathbf{e}$ (see Lindgren (1976) for an elementary discussion of maximum likelihood estimation); others, such as Hartigan and Anscombe in the written discussion following the Box and Cox paper, have suggested approximate methods for estimating λ. Here, we shall assume that the errors follow a normal distribution and outline a procedure by which a usual regression program can be used to obtain the maximum likelihood estimates $\hat{\lambda}$, $\hat{\boldsymbol{\beta}}$, and $\hat{\sigma}^2$.

Suppose for a moment that we know λ. Given this knowledge, estimation of $\hat{\boldsymbol{\beta}}$ and computation of the residual sum of squares is immediate:

$$\hat{\boldsymbol{\beta}}_\lambda = (\mathbf{X}^T \mathbf{X})^{-1} \mathbf{X}^T \mathbf{Y}^\lambda \tag{6.9}$$

$$RSS_\lambda = (\mathbf{Y}^\lambda)^T (\mathbf{I} - \mathbf{V}) \mathbf{Y}^\lambda \tag{6.10}$$

These would be obtained from an ordinary linear regression program by regressing $\mathbf{Y}^\lambda$ on the independent variables.

Since λ is unknown, (6.10) can be computed for a range of reasonable values for λ. For each $\lambda \neq 0$, compute

$$L(\lambda) = \frac{n}{2} \ln(\lambda^2) - \frac{n}{2} \ln(RSS_\lambda) + (\lambda - 1) \sum_{i=1}^{n} \ln(y_i) \tag{6.11a}$$

and if $\lambda = 0$, compute

$$L(\lambda) = -\frac{n}{2}\ln(RSS_\lambda) - \sum_{i=1}^{n}\ln(y_i) \qquad (6.11b)$$

(Box and Cox define the power transformation in a more general fashion to avoid special handling of $\lambda = 0$.) One can show that the λ chosen to maximize (6.11a) and (6.11b) is the maximum likelihood estimate assuming normal errors. Note that simply choosing λ to maximize RSS_λ does not make sense because for each λ the residual sum of squares is measured in a different scale. Equation (6.11) converts each RSS_λ to a common scale.

Example 6.2   Tree data

The data in the first two columns of Table 6.5 give the DBH, diameters at breast height in inches, and the ages in years of $n = 27$ chestnut oak trees grown on a poor site (Chapman and Demeritt, 1936). We wish to derive an equation relating DBH(Y) to age (X). From the scatter plot of Figure 6.9, both moderate curvature and heterogeneity of variance are apparent, so that, for a simple regression model to be adequate, a transformation may be desirable. We

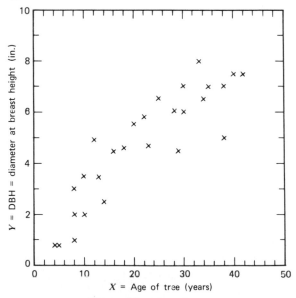

Figure 6.9   Tree data.

Table 6.5 Tree data

X = age	Y = DBH	−2	−1.5	−1.0	−0.5	0	0.5	1.0	1.5	2.0
						Y^λ for $\lambda =$				
4	0.8	1.5625	1.3975	1.250	1.118	−0.2231	0.8944	0.800	0.7155	0.640
5	0.8	1.5625	1.3975	1.250	1.118	−0.2231	0.8944	0.800	0.7155	0.640
8	1.0	1.0000	1.0000	1.000	1.000	0.0000	1.000	1.000	1.000	1.00
8	2.0	0.2500	0.3536	0.5000	0.7071	0.6931	1.414	2.000	2.828	4.00
8	3.0	0.1111	0.1925	0.3333	0.5774	1.099	1.732	3.000	5.196	9.00
10	2.0	0.2500	0.3536	0.5000	0.7071	0.6931	1.414	2.000	2.828	4.00
10	3.5	0.0816	0.1527	0.2857	0.5345	1.253	1.871	3.500	6.548	12.2
12	4.9	0.0416	0.0922	0.2041	0.4518	1.589	2.214	4.900	10.85	24.0
13	3.5	0.0816	0.1527	0.2857	0.5345	1.253	1.871	3.500	6.548	12.2
14	2.5	0.1600	0.2530	0.4000	0.6325	0.9163	1.581	2.500	3.953	6.25
16	4.5	0.0494	0.1048	0.2222	0.4714	1.504	2.121	4.500	9.546	20.2
18	4.6	0.0473	0.1014	0.2174	0.4663	1.526	2.145	4.600	9.866	21.1
20	5.5	0.0331	0.0775	0.1818	0.4264	1.705	2.345	5.500	12.90	30.2
22	5.8	0.0297	0.0716	0.1724	0.4152	1.758	2.408	5.800	13.97	33.6
23	4.7	0.0453	0.0981	0.2128	0.4613	1.548	2.168	4.700	10.19	22.0
25	6.5	0.0237	0.0603	0.1538	0.3922	1.872	2.550	6.500	16.57	42.2
28	6.0	0.0278	0.0680	0.1667	0.4082	1.792	2.449	6.000	14.70	36.0
29	4.5	0.0494	0.1048	0.2222	0.4714	1.504	2.121	4.500	9.546	20.2
30	6.0	0.0278	0.0680	0.1667	0.4082	1.792	2.449	6.000	14.70	36.0
30	7.0	0.0204	0.0540	0.1429	0.3780	1.946	2.646	7.000	18.52	49.0
33	8.0	0.0156	0.0442	0.1250	0.3536	2.079	2.828	8.000	22.63	64.0
34	6.5	0.0237	0.0603	0.1538	0.3922	1.872	2.550	6.500	16.57	42.2
35	7.0	0.0204	0.0540	0.1429	0.3780	1.946	2.646	7.000	18.52	49.0
38	5.0	0.0400	0.0894	0.2000	0.4472	1.609	2.236	5.000	11.18	25.0
38	7.0	0.0204	0.0540	0.1429	0.3780	1.946	2.646	7.000	18.52	49.0
40	7.5	0.0178	0.0487	0.1333	0.3651	2.015	2.739	7.500	20.54	56.2
42	7.5	0.0178	0.0487	0.1333	0.3651	2.015	2.739	7.500	20.54	56.2

shall try the Box and Cox method. In columns 3 through 11 of Table 6.5, values of Y^λ are given for $\lambda = -2, -1.5, -1, -0.5, 0, 0.5, 1, 1.5$, and 2. All of the computations needed to compute $L(\lambda)$ can be obtained by the regression of one of these columns on age. Thus if $\lambda = -2$ (column 3 of Table 6.5), we compute the regression of $1/Y^2$ on age. The residual sum of squares is 3.30. The quantity $\sum \ln(y_i)$ is the sum of the numbers in column 7 of Table 6.5, $\sum \ln(y_i) = 37.48$. Then, from (6.11),

$$L(-2) = \frac{27}{2} \ln\left((-2)^2\right) - \frac{27}{2} \ln(3.30) + (-2 - 1)(37.48)$$
$$= 18.71 - 16.11 - 112.44$$
$$= -109.84$$

$L(\lambda)$ was computed for each λ shown in Table 6.5, and the results are plotted in Figure 6.10. A smooth curve was then drawn through the points (additional values of λ between 1 and 2 could also be computed to help find $\hat{\lambda}$, since $\hat{\lambda}$ must be in this interval). From the graph, we read off $\hat{\lambda} \cong 1.3$ and $L(\hat{\lambda}) \cong -43.5$. Clearly, $L(\lambda)$ is nearly

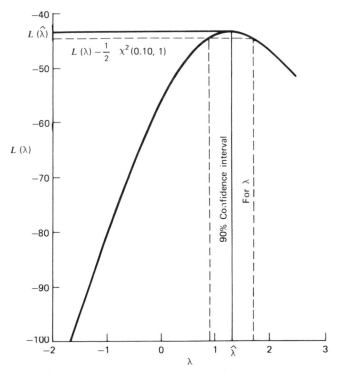

Figure 6.10 Log liklihood for tree data.

constant over the range from about 1 to 2, so any choice of λ in this range will nearly maximize the likelihood function. Using this idea and certain facts about the likelihood function, we can obtain a confidence interval for λ. A $(1 - \alpha) \times 100\%$ confidence interval for λ is the set of all points λ such that $L(\lambda) > L(\hat{\lambda}) - \frac{1}{2}\chi^2(\alpha, 1)$. For the example, a 90% confidence interval consists of all λ for which $L(\lambda) > -43.5 - \frac{1}{2}(2.71) = -44.9$. From Figure 6.10, we see this interval corresponds to the set $0.9 \leqslant \lambda \leqslant 1.7$. Since this interval includes the value $\lambda = 1$ corresponding to no transformation, it appears that little is to be gained by transforming Y. Box and Cox suggest using relatively simple choices for λ since the difference between, say, $\lambda = 1.3$ and $\lambda = 1.0$ is small in terms of the values of Y^λ but interpretation of the latter is generally simpler than interpretation of the former.

Transformation of the X's. Suppose we have p independent variables $X_1, X_2, \ldots, X_p$, and we contemplate transforming each by a power transformation to get $W_1, W_2, \ldots, W_p$, where, for $j = 1, 2, \ldots, p$,

$$W_j = \begin{cases} X_j^{\alpha_j} & \alpha_j \neq 0 \\ \ln(X_j) & \alpha_j = 0 \end{cases} \tag{6.12}$$

We will then fit the linear model

$$Y = \beta_0 + \sum \beta_j W_j + e \tag{6.13}$$

In (6.12) and (6.13) there is no requirement that the α_j's all be equal. Since, for all $\alpha = (\alpha_1, \ldots, \alpha_p)$, the dependent variable in (6.13) is the same, we can find an $\hat{\alpha}$ and a $\hat{\beta}$ that are (nonlinear) least squares estimators. This suggests that for each α we can fit model (6.13), get the residual sum of squares RSS_α, and then find $\hat{\alpha}$ to be the choice of α that minimizes RSS_α.

An iterative procedure for finding $\hat{\alpha}$ defined in this way has been provided by Box and Tidwell (1962). They begin with an initial guess of α, usually $\hat{\alpha}_1 = \hat{\alpha}_2 = \cdots = \hat{\alpha}_p = 1$. Then, by fitting an augmented linear model, they obtain an improved estimate of α, repeating this process until a desired level of convergence is reached.

To be specific, begin with $\hat{\alpha}_1 = \cdots = \hat{\alpha}_p = 1$, so $W_j = X_j^{\hat{\alpha}_j} = X_j$ and fit the regression model (6.13) to get

$$Y = \hat{\beta}_0 + \sum \hat{\beta}_j W_j \tag{6.14}$$

Next, create p new variables $Z_1, Z_2, \ldots, Z_p$, defined by

$$Z_j = W_j \ln(W_j) \qquad j = 1, 2, \ldots, p \tag{6.15}$$

The Z_j's are defined in such a way that, in the fitted augmented linear model,

$$\hat{y} = \hat{\beta}_0^* + \sum \hat{\beta}_j^* W_j + \sum \hat{\gamma}_j Z_j \tag{6.16}$$

each $\hat{\gamma}_j$ will be large (ignoring sign) if a transformation is needed, and small otherwise. The $\hat{\beta}_j^*$'s are not the same as the $\hat{\beta}_j$'s from fitted model (6.14). Box and Tidwell show that an appropriate estimate of α_j is

$$\hat{\alpha}_j = \left(\frac{\hat{\gamma}_j}{\hat{\beta}_j} + 1 \right) \times (\text{current value of } \hat{\alpha}_j) \tag{6.17}$$

The above procedure, with the $\hat{\alpha}_j$'s obtained from (6.17), can be repeated until the decrease in residual sum of squares is sufficiently small. However, a single cycle is often adequate.

As an example, we will apply this procedure to the tree data. Since $p = 1$, there is only one α_j, which we will simply call α. As a first guess, set $\hat{\alpha} = 1$, so $W = X$, and $Z = X \ln(X)$. The regression of Y on W is

$$\hat{Y} = 1.150716 + 0.1627836 W$$

while the regression of Y on W, Z is

$$\hat{Y} = -1.995722 + 0.8977549 W - 0.1825343 Z$$

from which we compute

$$\hat{\alpha} = \frac{-0.1825343}{0.1627836} + 1 = -0.1213310$$

(Box and Tidwell suggest keeping as many digits as possible in intermediate calculations.) Thus as a second guess, we take $W = X^{-0.1213310}$ and $Z = W \ln(W)$, refit the two models, and obtain $(\hat{\gamma}_j / \hat{\beta}_j) + 1 = -0.8653598$ at the second iteration. The suggested transformation is

$$W = (X^{-0.123310})^{-0.8653598} = X^{0.1049950}$$

The results of this procedure for six iterations are given in Table 6.6. The final value is $\hat{\alpha} \cong 0.0926$. However, for most purposes it is adequate to note that $\hat{\alpha}$ is nearly zero, and a log transformation is suggested. This conclusion could have been reached after only one or two iterations. Also, note that RSS_α is essentially constant over the range $-0.12 < \alpha < 0.1$, at least to the accuracy in the data. Thus six steps of the algorithm are shown here more as an exercise than as a practical procedure for analysis of the tree data.

The scatter plot of Y versus $\ln(X)$ is shown in Figure 6.11, demonstrating the success of this transformation, at least for trees 4 or more years old. However, the fitted model summarized in Table 6.7 would suggest that

Table 6.6 Six iterations for estimating α

Value of $\hat{\alpha}$	$\hat{\gamma}/\hat{\beta} + 1$	RSS_α
1.000000	-0.1213310	26.058875
-0.1213310	-0.8653598	19.679858
0.1049950	0.8723785	19.154400
0.0915954	1.0120028	19.152725
0.0926948	0.9990458	19.152713
0.0926063	1.0000775	19.152713
0.0926135		

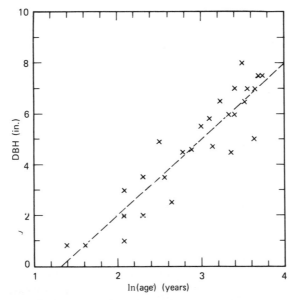

Figure 6.11 Scatter plot of $Y = $ DBH versus $X = $ ln(age) for the tree data (dotted line is the least squares line).

Table 6.7 Regression summary: Y on $\ln(X)$

Variable	Estimate	Standard Error	t-Value
Intercept	-3.928	0.762	-5.15
Slope	2.980	0.256	11.64

$R^2 = 0.844$, $\hat{\sigma}^2 = 0.802$, d.f. $= 24$

143

very young trees would have negative DBH, an impossible result. This does not invalidate the usefulness of a fitted model over the range of the data, but it does point out the problems of extrapolating from such a model. Also, note that in Table 6.7 a degree of freedom has been subtracted from the residual to reflect the fact that α has been estimated. In multiple regression, the degrees of freedom for error would be $n - p' - p$.

Andrews (1971) has suggested an overall test of the desirability of a transformation of the X's similar in spirit to Tukey's (1949) one degree of freedom test for nonadditivity in two-way tables. If $\hat{Y}$ is the vector of fitted values from the fit of the model $Y = X\beta + e$, define Z to be an $n \times 1$ vector whose ith element is $z_i = \hat{y}_i \ln(\hat{y}_i)$. Then, in the model $Y = X\beta + Z\gamma$, a t-test of $\gamma = 0$ provides a test of the need of a transformation. In the tree data, this test is $t = 3.02$ with 24 d.f., providing evidence that $\gamma \neq 0$, and suggesting the need for the transformation. After the first step in the iteration described above, $t = -1.51$, suggesting little need for transformation beyond this point. (Atkinson (1973) provides an analogous test for the need to transform Y.)

Problems

6.1 For Hooker's data (Problem 1.2), regress PRES on TEMP and log(PRES) on TEMP, and in each case graph r_i versus $\hat{y}_i$. Which model fits better? How and why?

6.2 For an appropriate part of the Berkeley Guidance Study (Problem 2.1), perform a complete case analysis. Are any transformations required for analysis? For the boys, make rankit plots for WT9 and for WT18. What do you conclude from these plots?

6.3 For the Longley data (Table 3.6) and the model Y on $X_1 X_2 X_3 X_4 X_5 X_6$, perform a case analysis. Are any of the D_i's large? What happens if the case with the largest D_i is deleted? Are there any apparent outliers?

6.4 The following data were collected in a study of the effect of dissolved sulfur on the surface tension of liquid copper (Baes and Kellogg, 1953):

X = Weight % Sulfur	Y = Decrease in Surface Tension (dynes/cm), 2 Replicates	
0.034	301	316
0.093	430	422
0.30	593	586
0.40	630	618
0.61	656	642
0.83	740	714

6.4.1. Find transformations of X and Y so that in the transformed scale the regression is linear.

6.4.2. Assuming that X is transformed to $\ln(X)$, which choice of Y gives better results, Y or $\ln(Y)$? (Sclove, 1972).

6.5 The following (hypothetical) data give stopping times for $n = 62$ trials of various automobiles traveling at speed X (miles per hour) and the resulting stopping distances Y (feet) (Ezekiel and Fox, 1959).

X	Y	X	Y
4	4	20	48
5	2, 4, 8, 8	21	39, 42, 55
7	7, 7	24	56
8	8, 9, 11, 13	25	33, 48, 56, 59
9	5, 5, 13	26	39, 41
10	8, 14, 17	27	57, 78
12	11, 19, 21	28	64, 84
13	15, 18, 27	29	54, 68
14	14, 16	30	60, 67, 101
15	16	31	77
16	14, 19, 34	35	85, 107
17	22, 29	36	79
18	29, 34, 47	39	138
19	30	40	110, 134

6.5.1. Draw the scatter plot Y versus X. Fit the simple regression model and draw the plot of r_i versus $\hat{y}_i$. What problems are apparent? Compute the F-test for lack of fit (Section 4.3) for the fitted model.

6.5.2. Use the Box and Tidwell method to find a transformation for X. Repeat 6.5.1.

6.5.3. Hald (1960) has suggested on the basis of a theoretical argument that the model $Y = \beta_1 X + \beta_2 X^2 + e$, with $\text{var}(e) = \sigma^2 X^2$ is appropriate for data of this type. Compare the fit of this model to the model fit in 6.5.2.

That is, for X in the range 0 to 40 mph, draw the curves that give the predicted values of Y from each model and qualitatively compare them. Also, for selected values of X, compute the variances of the predicted values. Then, for extrapolations to $X = 60$ mph, compute predicted values and their standard errors and compare. This comparison should demonstrate that an empirical relationship like that found in 6.5.2 may be perfectly adequate for some purposes such as interpolatory predictions, but may give completely misleading results for others, such as extrapolations or interpretation of parameter estimates (Draper and Hunter, 1969).

Table 6.7   Gas vapor data

X_1	X_2	X_3	X_4	Y	X_5
33.	53.	3.32	3.42	29.	30.
31.	36.	3.10	3.26	24.	24.
33.	51.	3.18	3.18	26.	26.
37.	51.	3.39	3.08	22.	22.
36.	54.	3.20	3.41	27.	27.
35.	35.	3.03	3.03	21.	20.
59.	56.	4.78	4.57	33.	32.
60.	60.	4.72	4.72	34.	34.
59.	60.	4.60	4.41	32.	27.
60.	60.	4.53	4.53	34.	34.
34.	35.	2.90	2.95	20.	19.
60.	59.	4.40	4.36	36.	37.
60.	62.	4.31	4.42	34.	34.
60.	36.	4.27	3.94	23.	22.
62.	38.	4.41	3.49	24.	24.
62.	61.	4.39	4.39	32.	32.
90.	64.	7.32	6.70	40.	38.
90.	60.	7.32	7.20	46.	44.
92.	92.	7.45	7.45	55.	57.
91.	92.	7.27	7.26	52.	53.
61.	62.	3.91	4.08	29.	29.
59.	42.	3.75	3.45	22.	21.
88.	65.	6.48	5.80	31.	28.
91.	89.	6.70	6.60	45.	45.
63.	62.	4.30	4.30	37.	39.
60.	61.	4.02	4.10	37.	39.
60.	62.	4.02	3.89	33.	34.
59.	62.	3.98	4.02	27.	26.
59.	62.	4.39	4.53	34.	34.
37.	35.	2.75	2.64	19.	19.
35.	35.	2.59	2.59	16.	15.
37.	37.	2.73	2.59	22.	22.

6.6 (John Rice). When gasoline is pumped into a tank, hydrocarbon vapors are forced out of the tank and into the atmosphere. To reduce this significant source of air pollution, devices are installed to capture the vapor. In testing these vapor recovery systems, the amount that escapes cannot be measured, but a "sniffer" can determine if some vapor is escaping. Also, the amount that is recovered can be measured. Thus to estimate the efficiency of the system, some method of estimating the total amount given off must be employed.

To this end, a laboratory experiment was conducted in which the amount of vapor given off was measured under carefully controlled conditions. There were four variables thought to be relevant for prediction:

X_1 = initial tank temperature, °F.

X_2 = temperature of the dispensed gasoline, °F.

X_3 = initial vapor pressure in the tank, pounds per square inch.

X_4 = vapor pressure of the dispensed gasoline, pounds per square inch.

In a sequence of 32 experimental fills, these conditions were varied and the quantity of emitted hydrocarbons, Y, was measured in grams. Also, the investigators developed, independently of the data, a theoretical model to predict hydrocarbons. The predictions from the theoretical model are included as X_5 in Table 6.7.

Analyze these data to find an empirical model for Y as a function of X_1, X_2, X_3, X_4. You should be concerned about necessary scale changes, outliers, and influential cases. In addition, how can the data be used to check the theoretical predictions given in X_5?

6.7 (Douglas Tiffany). To investigate the rent structure of Minnesota agricultural land with emphasis on alfalfa hay, each of the following variables was measured in 67 of the 87 counties in Minnesota (Table 6.8):

Y = cash rent collected per acre for alfalfa hay, in dollars.

X_1 = average rent per tillable acre, in dollars.

X_2 = number of dairy cows per square mile.

X_3 = total pasture acres ÷ total cropland acres.

Assume that the counties not measured have no alfalfa hay produced on rented land.

Analyze these data with a view toward investigating the relationship(s) between Y and the other variables. In particular, the investigator was interested in the amount of variation in Y that can be reasonably explained by the other variables.

Table 6.8　Land rent data

Y	X_1	X_2	X_3
18.38	15.50	17.25	0.24
20.00	22.29	18.51	0.20
11.50	12.36	11.13	0.12
25.00	31.84	5.54	0.12
52.50	83.90	5.44	0.04
82.50	72.25	20.37	0.05
25.00	27.14	31.20	0.27
30.67	40.41	4.29	0.10
12.00	12.42	8.69	0.41
61.25	69.42	6.63	0.04
60.00	48.46	27.40	0.12
57.50	69.00	31.23	0.08
31.00	26.09	28.50	0.21
60.00	62.83	29.98	0.17
72.50	77.06	13.59	0.05
60.33	58.83	45.46	0.16
49.75	59.48	35.90	0.32
8.50	9.00	8.89	0.08
36.50	20.64	23.81	0.24
60.00	81.40	4.54	0.05
16.25	18.92	29.62	0.72
50.00	50.32	21.36	0.19
11.50	21.33	1.53	0.10
35.00	46.85	5.42	0.08
75.00	65.94	22.10	0.09
31.56	38.68	14.55	0.17
48.50	51.19	7.59	0.13
77.50	59.42	49.86	0.13
21.67	24.64	11.46	0.21
19.75	26.94	2.48	0.10
56.00	46.20	31.62	0.26
25.00	26.86	53.73	0.43
40.00	20.00	40.18	0.56
56.67	62.52	15.89	0.05
51.79	56.00	14.25	0.15
96.67	71.41	21.37	0.05
50.83	65.00	13.24	0.08
34.33	36.28	5.85	0.10
48.75	59.88	32.99	0.21

Table 6.8 (continued)

Y	X_1	X_2	X_3
25.80	23.62	28.89	0.24
20.00	24.20	6.29	0.06
16.00	17.09	33.34	0.66
48.67	44.56	16.70	0.15
20.78	34.46	4.20	0.03
32.50	31.55	23.47	0.19
19.00	26.94	8.28	0.10
51.50	58.71	7.40	0.04
49.17	65.74	7.71	0.02
85.00	69.05	46.18	0.22
58.75	57.54	14.98	0.11
19.33	21.73	6.58	0.06
5.00	6.17	13.68	0.18
65.00	51.00	50.50	0.24
20.00	18.25	16.12	0.32
62.50	69.88	31.48	0.07
35.00	26.68	58.60	0.23
99.17	75.73	35.43	0.05
40.25	41.77	4.53	0.08
39.17	48.50	6.82	0.08
37.50	21.89	43.70	0.36
26.25	38.33	2.83	0.04
52.14	53.95	42.54	0.25
22.50	17.17	24.16	0.36
90.00	82.00	7.89	0.03
28.00	40.60	3.27	0.02
50.00	53.89	53.16	0.24
24.50	54.17	5.57	0.06

7

MODEL BUILDING I: DEFINING NEW VARIABLES

7.1 Polynomial regression

When the relationship between an independent variable X and a response Y is smooth, but not a straight line, linear regression models can be used if there is a scaling for X and Y so that the relationship in the transformed scale is straight. Alternatively, the model can be expanded by addition of terms that are powers of X, since any smooth function can be approximated by a polynomial of high enough degree. The resulting model is then given by

$$Y = \beta_0 + \beta_1 X + \beta_2 X^2 + \cdots + \beta_d X^d + \text{error} \qquad (7.1)$$

where d is the degree of the polynomial; if $d = 2$ the model is quadratic, $d = 3$ is cubic, and so on. In (7.1) we assume that the errors are independent with constant variance σ^2 (or else $\text{var}(e_i) = w_i \sigma^2$, with w_i known). When variance is not constant, a variance stabilizing transformation may be required before polynomials are fit. Polynomial models are usually fit as approximations as they almost never represent a physical model.

The model (7.1) can be analyzed via least squares by defining d new variables $Z_1, Z_2, \ldots, Z_d$ by $Z_1 = X$, $Z_2 = X^2, \ldots, Z_d = X^d$, so (7.1) can be written as

$$Y = \beta_0 + \beta_1 Z_1 + \cdots + \beta_d Z_d + \text{error} \qquad (7.2)$$

Then, any least squares program can in principle compute estimates of the

150

β's and the other usual regression statistics by the regression of Y on the Z's. However, if d is large (say $d = 3$ or more), serious numerical problems may arise and direct fitting of (7.2) can be unreliable. Details of better fitting methodology for large d are surveyed by Seber (1977, Chapter 8).

An example of quadratic regression has already been given with the physics data of Example 4.2. There, a test for lack of fit indicated that a straight-line model was not adequate for the data, while the test for lack of fit after the quadratic model indicated that this model was adequate. When a test for lack of fit is not available, comparison of the fit of the quadratic model

$$Y = \beta_0 + \beta_1 X + \beta_2 X^2 + \text{error} \tag{7.3}$$

to the simple linear regression model

$$Y = \beta_0 + \beta_1 X + \text{error} \tag{7.4}$$

is usually based on a t-test of $\beta_2 = 0$ in (7.3). Indeed, a strategy for choosing d is to continue adding terms to a model until the t-test for the highest order term is nonsignificant. Alternatively, an elimination scheme can be used, in which a maximum value of d is fixed, and terms are deleted from the model one at a time, starting with the highest order term, until the highest order remaining term has a significant t-value. Kennedy and Bancroft (1971) suggest using a significance level of about 0.10 for this procedure. However, in the most common uses of polynomial regression, it is enough to consider only $d = 1$ or $d = 2$. For larger values of d, the fitted polynomial curve becomes wiggly, providing an increasingly better fit by matching the variation in the observed data more and more closely. This curve is then modeling the random variation rather than the overall shape of the relationship between variables.

Polynomials with several independent variables. The extension to polynomials in several variables is straightforward: for each polynomial term added, a new independent variable is created. This also raises the possibility of models with cross-product terms that depend on several independent variables. For example, a model of the form

$$Y = \beta_0 + \beta_1 X_1 + \beta_2 X_2 + \beta_{11} X_1^2 + \beta_{22} X_2^2 + \beta_{12} X_1 X_2 + \text{error} \tag{7.5}$$

might be contemplated with two independent variables. To help understand the term $\beta_{12} X_1 X_2$, suppose first that the β's are known and $\beta_{12} = 0$. If X_1 is changed to $X_1 + \delta$ then the response Y will change from (7.5) to Y', where

$$Y' = \beta_0 + \beta_1(X_1 + \delta) + \beta_{11}(X_1 + \delta)^2 + \beta_2 X_2 + \beta_{22} X_2^2 \tag{7.6}$$

and the change in Y is

$$\Delta Y = Y' - Y = \beta_1 \delta + \beta_{11}(2X_1\delta + \delta^2) \qquad (7.7)$$

so ΔY does not depend on X_2, although it does depend on X_1. Now, if $\beta_{12} \neq 0$, and if X_1 is changed to $X_1 + \delta$, the change in Y is

$$\Delta Y = \beta_1\delta + \beta_{11}(2X_1\delta + \delta^2) + \beta_{12}\delta X_2 \qquad (7.8)$$

and the effect on Y due to altering X_1 will depend on the value of X_2 as well as the value of X_1. Thus if $\beta_{12} \neq 0$, an interaction effect between X_1 and X_2 is modeled.

Response surfaces. Experimental designs to estimate parameters of a polynomial, perhaps to find combinations of X's to produce a maximum or minimum for Y, are called response surface designs. Discussions of them are given by Box and Wilson (1951), John (1971), Myers (1971), and Box et al. (1978).

7.2   Dummy variables: dichotomous

Dummy variables (or indicators) are used to include categorical independent variables in a regression analysis. Many of these variables may have two categories, such as sex (male or female), treatment condition (treatment or no treatment), and presence of disease or symptom (yes or no). They usually take on the two values zero and one to indicate which category is the correct one for a given case; assignment of labels to the values is generally arbitrary.

Example 7.1   Estimating the effect of cloud seeding

Judging the success or failure of experiments that are intended to modify weather, such as cloud seeding experiments with the purpose of increasing rainfall, is an important statistical problem. Results from these experiments have generally been mixed: sometimes the observed effect of seeding has been to increase rainfall, sometimes to decrease rainfall, and sometimes no effect is observed. Presumably, these mixed results are due at least in part to latent variation not directly controlled or measured in the experiment. Furthermore, the effects of seeding may result in only a relatively small average change in the total amount of rain that falls relative to the natural variation in rainfall; however, these small changes can have enormous consequences in the overall amount of moisture that is produced.

The data given in Table 7.1 are taken from the Florida Area Cumulus Experiment (FACE) collected in 1975 (Woodley et al., 1977). A simplified version of the experimental protocol is as follows. A fixed target area of approximately 3000 square miles was established to the north and east of Coral Gables, Florida. During the summer of 1975, each day was judged on its suitability for seeding. The decision to use a particular day in the experiment was based primarily on a suitability criterion S, where S is a computed number depending on a mathematical model for rainfall. Days with $S \geqslant 1.5$ were chosen as experimental days; there were 24 days chosen in 1975. On each day, the decision to seed was made by flipping a coin; as it turned out, 12 days were seeded, 12 unseeded. On seeded days, silver iodide was injected into the clouds from small aircraft. In all, for each experimental day, the following quantities were measured:

A = action (1 = seed, 0 = do not seed).

D = days after the first day of the experiment (June 16, 1975 = 0).

S = suitability for seeding.

C = percent cloud cover in the experimental area, measured using radar in Coral Gables, Florida.

P = prewetness, amount of rainfall in the hour preceding seeding, in 10^7 cubic meters.

E = echo motion category, either 1 or 2, a measure of the type of cloud.

Y = rainfall in hours following the action of seeding or not seeding, in 10^7 cubic meters.

The problem we shall study is estimating the effect on rainfall due to seeding. A secondary interest not considered here in any detail is modeling rainfall by the other measured variables. Randomization was used in this experiment because rainfall will be influenced by many other factors that have not been included in the study. One hopes that over the course of the experiment the effects of these other factors will balance, benefitting seeded and unseeded days equally. With this in mind, we are encouraged to consider a linear model for rainfall, even though the relationship between rainfall and the other predictors is probably much more complicated. As usual, we hope that a fitted linear model will provide a useful approximation to the true state of affairs.

Careful thought should be given to the possible nature of the effect due to the dummy variable A. This effect, if it exists, can manifest itself in at least two different ways. First, the effect of rainfall could

Table 7.1 Cloud seeding data

Case	A	D	S	C	P	Log(P)	E	SA	CA	PA	Log(P)A	EA	Y	Log(Y)
1	0	0	1.75	13.40	0.274	−0.56225	2	0	0	0	0	0	12.85	1.10890
2	1	1	2.70	37.90	1.267	0.10278	1	2.70	37.90	1.267	0.10278	1	5.52	0.74194
3	1	3	4.10	3.90	0.198	−0.70333	2	4.10	3.90	0.198	−0.70333	2	6.29	0.79865
4	0	4	2.35	5.30	0.526	−0.27901	1	0	0	0	0	0	6.11	0.78604
5	1	6	4.25	7.10	0.250	−0.60206	1	4.25	7.10	0.250	−0.60206	1	2.45	0.38917
6	0	9	1.60	6.90	0.018	−1.74473	2	0	0	0	0	0	3.61	0.55751
7	0	18	1.30	4.60	0.307	−0.51286	1	0	0	0	0	0	0.47	−0.32790
8	0	25	3.35	4.90	0.194	−0.71220	1	0	0	0	0	0	4.56	0.65896
9	0	27	2.85	12.10	0.751	−0.12436	1	0	0	0	0	0	6.35	0.80277
10	1	28	2.20	5.20	0.084	−1.07572	1	2.20	5.20	0.084	−1.07572	1	5.06	0.70415
11	1	29	4.40	4.10	0.236	−0.62709	1	4.40	4.10	0.236	−0.62709	1	2.76	0.44091
12	1	32	3.10	2.80	0.214	−0.66959	1	3.10	2.80	0.214	−0.66959	1	4.05	0.60746
13	0	33	3.95	6.80	0.796	−0.09909	1	0	0	0	0	0	5.74	0.75891
14	1	35	2.90	3.00	0.124	−0.90658	1	2.90	3.00	0.124	−0.90658	1	4.84	0.68485
15	1	38	2.05	7.00	0.144	−0.84164	1	2.05	7.00	0.144	−0.84164	1	11.86	1.07408
16	0	39	4.00	11.30	0.398	−0.40012	1	0	0	0	0	0	4.45	0.64836
17	0	53	3.35	4.20	0.237	−0.62525	2	0	0	0	0	0	3.66	0.56348
18	1	55	3.70	3.30	0.960	−0.01773	1	3.70	3.30	0.960	−0.01773	1	4.22	0.62531
19	0	56	3.80	2.20	0.230	−0.63827	1	0	0	0	0	0	1.16	0.06446
20	1	59	3.40	6.50	0.142	−0.84771	2	3.40	6.50	0.142	−0.84771	2	5.45	0.73640
21	1	65	3.15	3.10	0.073	−1.13668	1	3.15	3.10	0.073	−1.13668	1	2.02	0.30535
22	0	68	3.15	2.60	0.136	−0.86646	1	0	0	0	0	0	0.82	−0.08619
23	1	82	4.01	8.30	0.123	−0.91009	1	4.01	8.30	0.123	−0.91009	1	1.09	0.03743
24	0	83	4.65	7.40	0.168	−0.77469	1	0	0	0	0	0	0.28	−0.55284

be *additive*: For any set of combinations of other variables, the effect
of seeding the clouds will be to increase (or, perhaps decrease)
rainfall by some fixed amount, without depending on any other
variables. If this were the case, then one would have a model of the
form

$$Y = \beta_0 + \beta_1 A + \text{(a linear function of } D, S, C, E, \text{ and } P) + e \quad (7.9)$$

In the above equation, the linear function of the other predictors is
not specified exactly, since they may interact together or need to be
transformed, polynomial terms may be required, or some of the
predictors may be profitably dropped from the model. If this model
were appropriate for seeding experiments, then the effect of seeding
would be to change the expected rainfall by an amount β_1 cubic
meters ($\times 10^7$), regardless of the other predictors.

The meaning of this model may be seen graphically in Figure 7.1 for
a problem with a single additional predictor (or covariate). In this
graph, the x axis is the value of the covariate and the y axis is the
value of the response. Separate regression lines are drawn for $A = 0$
and $A = 1$. Under the additive treatment effect model, the two lines
are parallel.

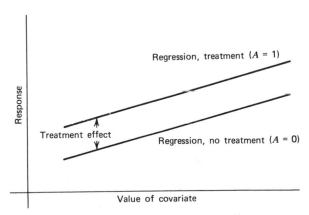

Figure 7.1   Additive treatment effect.

A second more comprehensive class of models for seeding would
allow that the effect of seeding may be different for different combinations of the other predictors. For example, it may be true that if
the seedability criterion S is small, then seeding has a small effect,
but if seedability S is large, then the effect of seeding is large. This is
shown in Figure 7.2, with nonparallel regression lines for each

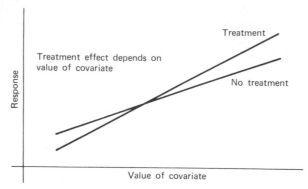

Figure 7.2   A nonadditive treatment effect.

treatment. Notice that in Figure 7.2 the treatment effect (the vertical distance from the treatment line to the no treatment line) may be of opposite signs for some values of the covariates, indicating that the seeding effect can be positive for some values of the covariates and negative for others.

The model for nonparallel regressions is more complicated than the simple additive treatment effect model. We can write it down in two nearly equivalent ways. First, we could model a separate equation for seeded ($A = 1$) and unseeded ($A = 0$) days. Using this approach, one would also obtain two separate estimates of error variance, one from each group. If we assume that the error variance is the same in each group, an alternative and simpler approach can be obtained by writing a single equation for all the data. This is done by defining a new set of predictors, which are products of the indicator for seeding (A) and the other predictors (S, C, E, and P). These four new variables are defined by $SA = S \times A$, $CA = C \times A$, $EA = E \times A$, $PA = P \times A$; they are given in Table 7.1. Note that, for example, the values of SA are those of S if $A = 1$ and 0 if $A = 0$. The full model is then given by

$$Y = \beta_0 + \beta_1 A + \beta_2 D + \beta_3 S + \beta_4 C + \beta_5 E + \beta_6 P$$
$$+ \beta_7 SA + \beta_8 CA + \beta_9 EA + \beta_{10} PA + e \qquad (7.10)$$

If the treatment effect is additive (does not depend on S, C, E, or P) then $\beta_7 = \beta_8 = \beta_9 = \beta_{10} = 0$. If there is no treatment effect at all, then $\beta_1 = \beta_7 = \beta_8 = \beta_9 = \beta_{10} = 0$. In an analysis where nonadditive treatment effects are suspected, (7.10) is a reasonable first model.

Because the analysis of these data is very complex, we shall pursue it in detail. The first consideration is to find appropriate scaling for the variables. We see that the variables P and Y are volume measures, while the other variables are linear. To transform to a common scale, and to insure positive estimates of rainfall, logarithmic transformations for Y and P might be entertained. This need is borne out by a plot of r_i versus $\hat{y}_i$ or r_i versus S for the model (7.10), where a shape reminiscent of Figure 6.1g is obtained. Also, for some days in the experiment, the model (7.10) leads to negative values of $\hat{y}_i$. Both considerations suggest the following model rather than (7.10):

$$\log Y = \beta_0 + \beta_1 A + \beta_2 D + \beta_3 S + \beta_4 C + \beta_5 E + \beta_6(\log P)$$
$$+ \beta_7 SA + \beta_8 CA + \beta_9 EA + \beta_{10}(\log P)A + e \qquad (7.11)$$

In this model, the cloud seeding effect is now to be measured in units of $\log(10^7$ cubic meters) rather than cubic meters, which, for purposes of judging the existence of an effect, should be equally useful.

The case statistics for the fit of (7.11) are listed in Table 7.2. Three cases are of interest: case 2 because of $D_2 = 1.51$ and cases 7 and 24,

Table 7.2   Case statistics

Case	$\log(Y)$	$\hat{e}_i$	r_i	v_{ii}	D_i	t_i
1	1.1089	$-0.7352E-01$	-0.3986	0.5801	0.0200	-0.39
2	0.7419	$-0.3156E-01$	-0.6765	0.9731	1.5074	-0.66
3	0.7986	-0.1527	-0.8744	0.6236	0.1152	-0.87
4	0.7860	0.2190	0.9231	0.3053	0.0340	0.92
5	0.3892	-0.1776	-0.8068	0.4020	0.0398	-0.80
6	0.5575	0.1628	1.2538	0.7919	0.5437	1.28
7	-0.3279	-0.5165	-2.7333	0.5593	0.8620	-4.03
8	0.6590	0.3524	1.3884	0.2051	0.0452	1.45
9	0.8028	0.1329	0.5957	0.3853	0.0202	0.58
10	0.7041	-0.1876	-0.8310	0.3714	0.0371	-0.82
11	0.4409	$0.8024E-01$	0.3446	0.3307	0.0053	0.33
12	0.6075	-0.1339	-0.5197	0.1811	0.0054	-0.50
13	0.7589	0.1274	0.5291	0.2849	0.0101	0.51
14	0.6849	$-0.1780E-01$	-0.0687	0.1702	0.0001	-0.07
15	1.0741	0.1862	0.8119	0.3509	0.0324	0.80
16	0.6484	$0.8102E-01$	0.3570	0.3644	0.0066	0.34
17	0.5635	$-0.8931E-01$	-0.6639	0.7767	0.1394	-0.65
18	0.6253	$0.8705E-01$	0.7812	0.8468	0.3066	0.77
19	$0.6446E-01$	$0.1160E-01$	0.0474	0.2612	0.0001	0.05
20	0.7364	0.1527	0.8744	0.6236	0.1152	0.87
21	0.3054	$0.1420E-01$	0.0591	0.2889	0.0001	0.06
22	$-0.8619E-01$	0.1401	0.5877	0.2982	0.0133	0.57
23	$0.3743E-01$	0.1807	0.9694	0.5712	0.1138	0.97
24	-0.5528	-0.5479	-2.6047	0.4539	0.5126	-3.62

because of $t_7 = -4.03$ and $t_{24} = -3.62$. From the large value of D_2, we know that deletion of this case may result in substantial changes in the fitted model. Returning to Table 7.1, we note that on this day the cloud cover. C was 37.9%; the authors of the original paper classify this as a "disturbed day," and it is clearly different from the other days in the study, so exclusion of case 2 from the data before further analysis may be a reasonable procedure. However, this is not a general rule, and deletion of influential cases should only be done where there is reason to believe that the case is really different from the rest of those in the data. Both cases 7 and 24 have very large t_i, each with p-value near 0.05 for the outlier test. Referring to the original data, we see that both these days are unseeded, and have very low observed rainfalls. From their associated v_{ii}, neither case is particularly unusual, nor from their D_i are they overly influential, at least for model (7.11). However, because of the unusually small observed rainfall, we shall delete them before further analysis. A thorough analysis would require repeating the analysis with them included (this point is pursued for these data by Cook and Weisberg (1980)).

Thus before proceeding, we delete cases 2, 7, and 24, leaving $n = 21$ cases remaining. The residuals from the fit of model (7.11) are plotted in Figure 7.3 and the regression summaries are given in Table 7.3. The plot and summaries indicate no clear violation of assumptions.

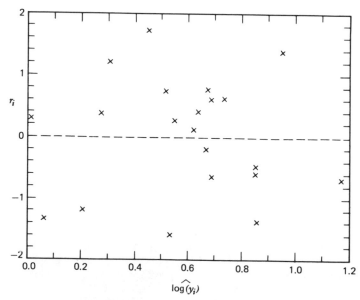

Figure 7.3 Residual plot, full model, cases 2, 7, and 24 deleted.

Table 7.3   Regression full model with log(Y) as dependent variable

Variable	Estimate	Standard Error	t-value
Intercept	0.417091	0.399500	1.04
A	1.42629	0.510010	2.80
D	$-0.625920E-02$	$0.168467E-02$	-3.72
S	$0.598899E-02$	$0.849820E-01$	0.07
C	$0.300897E-01$	$0.150398E-01$	2.00
log(P)	0.341399	0.145797	2.34
E	0.265071	0.135283	1.96
SA	-0.333451	0.107271	-3.11
CA	$-0.228671E-01$	$0.281945E-01$	-0.81
log(P)A	$0.729695E-01$	0.223604	0.33
EA	$0.499091E-01$	0.178301	0.28

$\hat{\sigma}^2 = 0.193871E-01$, d.f. $= 10$, $R^2 = 0.90$

Table 7.4   Final model for log(Y)

Variable	Estimate	Standard Error	t-value
Intercept	0.492488	0.129108	3.81
A	1.29442	0.190071	6.81
D	$-0.663505E-02$	$0.126249E-02$	-5.26
C	$0.218656E-01$	$0.966504E-02$	2.26
log(P)	0.399014	$0.830460E-01$	4.80
E	0.300956	$0.735401E-01$	4.09
SA	-0.326316	$0.519781E-01$	-6.28

$\hat{\sigma}^2 = 0.148910E - 01$, d.f. $= 14$, $R^2 = 0.89$

However, the small t-values in Table 7.3 indicate that the fitted model can be improved by deleting some of the variables. Using methodology for variable selection that will be developed in the next chapter, we are led to the model summarized in Table 7.4. The fitted model is

$$\widehat{\log Y} = 0.492 + 1.294A - 0.007D + 0.022C + 0.399(\log P)$$
$$+ 301E - 0.326SA \qquad (7.12)$$

Note that the coefficient for SA is nonzero, and that the effect of seeding depends on S, so we are in a situation like that of Figure 7.2.

To explore this dependence, the expected change in $\widehat{\log(Y)}$ if an unseeded day were seeded is

$$\Delta(\widehat{\log Y}) = \left[(\widehat{\log Y}) \text{ if } A = 1\right] - \left[(\widehat{\log Y}) \text{ if } A = 0\right]$$
$$= \hat{\beta}_1 + \hat{\beta}_7(SA) = 1.294 - 0.326S \qquad (7.13)$$

Figure 7.4 is a graph of the change in log(rainfall) as a function of the seedability criterion S. We have the curious result that as seeding suitability increases, the added rainfall appears to decrease, and becomes negative if $S > 4$. The vertical distance between the two curved lines in Figure 7.4 gives the 95% confidence interval for $\Delta(\widehat{\log Y})$. To compute this, first find the variance of $\Delta(\widehat{\log Y})$,

$$\text{var}(\Delta(\widehat{\log Y})) = \text{var}(\hat{\beta}_1) + S^2 \text{var}(\hat{\beta}_7) - 2S \, \text{cov}(\hat{\beta}_1, \hat{\beta}_7) \quad (7.14)$$

The methodology for estimating each of these variances and covariances to obtain $\text{var}(\Delta \widehat{\log Y})$ is discussed in Chapter 2. The 95% confidence bounds are then found in analogy to (1.38).

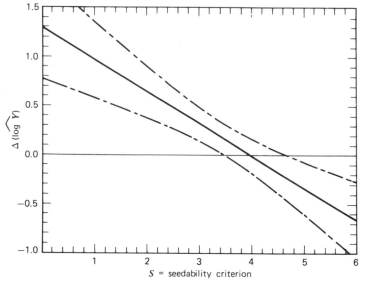

Figure 7.4   Added log(rainfall) as a function of seedability.

7.3   Dummy variables: polytomous

If a categorical variable has t possible values, then $t - 1$ dummy variables are required to fit an additive effect and even more are required to fit a nonadditive effect. Furthermore, the dummy variables can be defined in many ways although the results obtained are generally equivalent.

Suppose, in the example of the last section, three treatments were used, namely, unseeded, seeded with silver iodide, and seeded with some other agent such as dry ice. We could define two dummy variables

$$A_1 = 1 \quad \text{if unseeded, 0 otherwise}$$

$$A_2 = 1 \quad \text{if silver iodide, 0 otherwise}$$

Then, $A_1 = A_2 = 0$ only if seeding with dry ice was carried out. If we consider only one other variable, say S, the additive (parallel regression) model is

$$Y_0 = \beta_0 + \beta_1 A_1 + \beta_2 A_2 + \beta_3 S + e \tag{7.15}$$

The regressions are parallel because, for each group β_3 is the same, and the intercepts for the three groups are β_0 for dry ice, $\beta_0 + \beta_1$ for unseeded and $\beta_0 + \beta_2$ for silver iodide. The F-test of $\beta_1 = \beta_2 = 0$ (β_0, β_3 arbitrary) tests for the equality of intercepts for the three treatment conditions.

As pointed out, the definition of the dummy variables is not unique, since many other alternatives are possible such as defining A_1' and A_2' as follows:

Unseeded days: $\quad A_1' = 1 \qquad A_2' = 0$

Silver iodide: $\qquad A_1' = 0 \qquad A_2' = 1$

Dry ice: $\qquad\quad A_1' = -1 \qquad A_2' = -1$

In the model

$$y = \beta_0 + \beta_1' A_1' + \beta_2' A_2' + \beta_3 S + e \tag{7.16}$$

the F-test for $\beta_1' = \beta_2' = 0$ given β_0, β_3 will be identical to the obtained F from (7.15).

Since the definition of dummy variables is arbitrary, some care is required in defining treatment effects. Usually, comparing two treatments will be of interest. This is done by taking the difference between the fitted responses. For example, the estimated difference of silver iodide and control is, in model (7.15),

(predicted value given silver iodide) − (predicted value given control)

$$= (\hat{\beta}_0 + \hat{\beta}_2 + \hat{\beta}_3 S) - (\hat{\beta}_0 + \hat{\beta}_1 + \hat{\beta}_3 S) = \hat{\beta}_2 - \hat{\beta}_1$$

$$\tag{7.17}$$

while the estimated difference between silver iodide and dry ice is

(predicted value given silver iodide) − (predicted value given dry ice)

$$= (\hat{\beta}_0 + \hat{\beta}_2 + \hat{\beta}_3 S) - (\hat{\beta}_0 + \hat{\beta}_3 S) = \hat{\beta}_2$$

$$\tag{7.18}$$

Occasionally some researchers define a single dummy variable with more than two values to represent a polytomous categorical variable. For example, one could at least in principle define a variable taking on the value 0 for control days, 1 for silver iodide, and 2 for dry ice. In general, this practice should be avoided since it assumes that the categories are ordered as given (why should dry ice get a higher numbering than silver iodide?), and that the difference between any two groups is exactly the same (changing from the control to the silver iodide condition has exactly the same outcome as changing from dry ice to silver iodide). These assumptions are rarely valid.

Many dummy variables. Regression equations that include many dummy variables occur regularly. For example, a study of voting behavior might include dummy variables for sex (two categories), party affiliation (three categories), age (categorized into several age classes), place of residence (perhaps categorized by size of town, or by neighborhood, etc.), political beliefs (liberal, conservative, etc.), and several others. If there are many dummy variables, fitting separate regressions to each group of cases with common values on all of the dummy variables will often be impossible because each group will be very small. The analyst is forced to assume the parallel regression model for most, or all, of the dummy variables in which the effect of each of the categorizations is additive. The resulting regression analysis will only be as reliable as is the assumption of parallel regressions.

7.4   Comparing regression lines

For simplicity, consider simple regression with m groups of cases, such that, in the kth group with n_k cases, the correct model is

$$Y = \beta_{0k} + \beta_{1k}X + \text{error} \qquad k = 1, 2, \ldots, m$$

Comparison of these m regression lines, based on the observed data, is often of interest. We distinguish four different situations:

Model 1   Most general. If all the parameters are different, we have a circumstance like that in Figure 7.5a.

Model 2   Parallel regressions. In this model, $\beta_{11} = \beta_{12} = \cdots = \beta_{1m}$, but the intercepts are arbitrary. These are the circumstances that lead to the additive model for dummy variables, as in Section 7.2 and shown in Figure 7.5b.

Model 3   Concurrent regressions. In this model the intercepts are all equal, $\beta_{01} = \cdots = \beta_{0m}$, but slopes are arbitrary, as illustrated in Figure

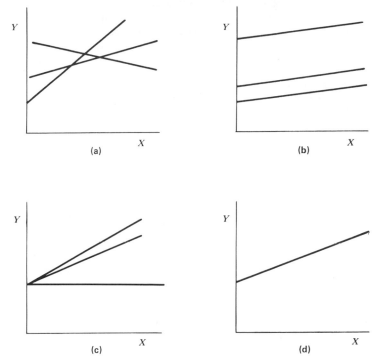

Figure 7.5   Four models for comparing regressions. **(a)** Most general. **(b)** Parallel. **(c)** Concurrent. **(d)** Coincident.

7.5c. The more general model, where the regression lines are concurrent at an arbitrary point, is more difficult, and is discussed by Seber (1977) and Saw (1966).

Model 4   Coincident regression lines. Here, all lines are the same, $\beta_{01} = \cdots = \beta_{0m}$ and $\beta_{11} = \cdots = \beta_{1m}$. This is the most stringent model, as illustrated in Figure 7.5d.

It is usually of interest to test the plausibility of models 4 or 2 against a different, less stringent model as an alternative. The form of these tests is immediate from the formulation of the general F-test given in Section 4.4. The methodology can be best illustrated by an example.

Example 7.2   Twin data

The data in Table 7.5 give the presumed IQ scores of identical twins, one raised in a foster home (Y), and the other raised by natural parents (X). The data were originally used by Burt (1966). For the purposes of an example, we can divide cases into three groups according to the social class of the natural parents.

Table 7.5　Twin data for comparing regression lines

Case Number	Y	X	W_1	W_2	W_3	Z_1	Z_2	Z_3
1	82	82	1	0	0	82	0	0
2	80	90	1	0	0	90	0	0
3	88	91	1	0	0	91	0	0
4	108	115	1	0	0	115	0	0
5	116	115	1	0	0	115	0	0
6	117	129	1	0	0	129	0	0
7	132	131	1	0	0	131	0	0
8	71	78	0	1	0	0	78	0
9	75	79	0	1	0	0	79	0
10	93	82	0	1	0	0	82	0
11	95	97	0	1	0	0	97	0
12	88	100	0	1	0	0	100	0
13	111	107	0	1	0	0	107	0
14	63	68	0	0	1	0	0	68
15	77	73	0	0	1	0	0	73
16	86	81	0	0	1	0	0	81
17	83	85	0	0	1	0	0	85
18	93	87	0	0	1	0	0	87
19	97	87	0	0	1	0	0	87
20	87	93	0	0	1	0	0	93
21	94	94	0	0	1	0	0	94
22	96	95	0	0	1	0	0	95
23	112	97	0	0	1	0	0	97
24	113	97	0	0	1	0	0	97
25	106	103	0	0	1	0	0	103
26	107	106	0	0	1	0	0	106
27	98	111	0	0	1	0	0	111

For the example, $m = 3$, $n_1 = 7$, $n_2 = 6$, and $n_3 = 14$, giving $n = \sum n_k = 27$. The data are graphed in Figure 7.6, suggesting that the regression lines may well be the same. For the tests, we treat the IQ of the twin raised at home as a standard to which the twin from the foster home is to be compared, to study the effect of being raised apart. In addition to comparing the different lines for different social classes, one might also be interested in the actual values of the estimated slope, and whether the slope differs from 1. This point will not be pursued here.

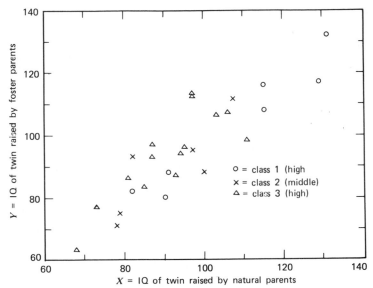

Figure 7.6 Data for twins.

In addition to X and Y, six other variables are given, namely, three dummy variables W_1, W_2, and W_3 to indicate social class ($W_1 = 1$ for highest class, $W_2 = 1$ for middle, $W_3 = 1$ for lowest), and three additional variables $Z_1 = W_1 X$, $Z_2 = W_2 X$, and $Z_3 = W_3 X$. These variables will aid in fitting the four models suggested for comparing regression lines.

Model 1. To fit Model 1, a separate regression calculation can be done for each group to obtain parameter estimates. The *RSS*, say RSS_1, is then obtained by adding the separate *RSS* from each regression. This *RSS* has $df_1 = \sum (n_k - p')$ degrees of freedom (for simple regression, $p' = 2$). Alternatively and equivalently, one could fit the model

$$Y = \beta_{01} W_1 + \beta_{02} W_2 + \beta_{03} W_3 + \beta_{11} Z_1 + \beta_{12} Z_2 + \beta_{13} Z_3 + e$$

with the W's and Z's as given in Table 7.5 (note that no overall intercept is in this model). The results of this fit are given in Table 7.6.

Table 7.6   Computations for twin data

	Estimates (t-Values)			
Variable	Model 1	Model 2	Model 3	Model 4
Intercept	—	—	2.56 (0.24)	9.21 (0.99)
X	—	0.97 (9.03)		0.90 (9.36)
W_1	$-1.87\ (-0.11)$	$-0.61\ (-0.05)$		
W_2	0.82 (0.03)	1.43 (0.14)		
W_3	7.20 (0.43)	5.62 (0.56)		
Z_1	0.98 (5.99)		0.94 (9.44)	
Z_2	0.97 (3.40)		0.95 (7.93)	
Z_3	0.95 (5.21)		1.00 (8.57)	
RSS	1317	1318	1326	1494
d.f.	21	23	23	25

Model 2. To fit parallel regressions, use the model

$$Y = \beta_{01} W_1 + \beta_{02} W_2 + \beta_{03} W_3 + \beta_1 X + \text{error}$$

to get estimates of each intercept and of the common slope β_1. The RSS from this fit, say RSS_2, has $\mathrm{df}_2 = \sum n_k - m - p$ degrees of freedom (for a simple regression $p = 1$). This fit is summarized in Table 7.6.

Model 3. This model requires fitting

$$Y = \beta_0 + \beta_{11} Z_1 + \beta_{12} Z_2 + \beta_{13} Z_3 + \text{error}$$

The RSS for this model, RSS_3, has $\mathrm{df}_3 = \sum n_k - mp - 1$ degrees of freedom, as shown in Table 7.6.

Model 4. This model assumes common regression lines, so the estimates are computed by pooling the data to fit

$$Y = \beta_0 + \beta_1 X + e$$

The RSS for this model, RSS_4, has $\mathrm{df}_4 = n - p'$ degrees of freedom as shown in Table 7.6.

Most tests concerning the slopes and intercepts of different regression lines will use the general model (model 1) as the alternative model. For example, testing for parallel lines will use the general model as the alternative. The usual F-test, for testing models 2, 3, and 4, is then given, for $l = 2, 3, 4$ by

$$F_l = \frac{(RSS_l - RSS_1)/(df_l - df_1)}{RSS_1/df_1}$$

$$l = 2, 3, 4 \text{ with } (df_l - df_1, df_1) \text{ d.f.} \qquad (7.19)$$

If the hypothesis provides as good a model as does the alternative, then F will be small. If the model is not adequate when compared to the general model, then F will be large (when compared to the percentage points of the $F(df_l - df_1, df_1)$ distribution).

For the twin data, the F-statistics are

$$F_2 = \frac{(1318 - 1317)/2}{1317/21} = 0.01$$

$$F_3 = \frac{(1326 - 1317)/2}{1317/21} = 0.08$$

$$F_4 = \frac{(1494 - 1317)/4}{1317/21} = 0.71$$

Since all these F values are much less than 1, comparison to percentage points of the corresponding F-distribution is unnecessary. The most restrictive model, model 4, is as good as the least restrictive model, model 1. The tiny value of $F_2 = 0.01$ may serve as a flag to the careful analyst, for, under the model assumed, observing an F-statistic this small is extremely unlikely (if all assumptions are true, the probability of F_2 being less than its observed value in this problem is less than 0.005). The observed data agree with the theory much more closely than chance theory would lead us to expect.

Analysis of covariance. Testing model 4 against model 2 is usually called the analysis of covariance. The independent variable is called a covariate and is expected to have the same effect in all groups. The difference between groups is assumed to be an additive treatment effect. These two assumptions combine to give the assumption that the correct alternative hypothesis is equal slopes in groups, but possibly different intercepts (the difference between intercepts is the "treatment effect").

7.5   Scaling of variables

In models with an intercept, least squares regression has a very pleasant property called location and scale invariance: if any of the variables in the data (independent or dependent) is scaled by addition of a constant or by multiplication by a constant, then in the resulting regression, estimates will be changed in a predictable way and scale-free quantities, such as R^2 and F- and t-tests will be unaffected. In this section, the effects of scaling variables are summarized. Let $\hat{\beta}_0, \hat{\beta}_1, \ldots, \hat{\beta}_p$ and $\hat{\sigma}^2$ be the least squares estimates obtained before scaling. Only the effects of scaling on these and on R^2 and test statistics are given. The effects on estimated standard errors are easily derived from the results given here.

Scaling independent variables. If a variable X_j is replaced by $(X_j - c_j)/d_j$, then, in the resulting regression $\hat{\beta}_j$ is increased to $d_j\hat{\beta}_j$, $\hat{\beta}_0$ becomes $\hat{\beta}_0 + c_j\hat{\beta}_j$, and $\hat{\sigma}^2$, F-tests, t-tests, and R^2 are unaffected.

Scaling the response. If Y is replaced by $(Y - f)/g$, then $\hat{\beta}_0$ is replaced by $(\hat{\beta}_0 - f)/g$, each $\hat{\beta}_j$ $(j = 1, \ldots, p)$ gets replaced by $\hat{\beta}_j/g$, and all sums of squares in the analysis of variance and $\hat{\sigma}^2$ are divided by g^2. However, F- and t-tests are unaffected.

A common use of scaling is to standardize the variables by subtracting off sample means and dividing by sample standard deviations. Thus for $j = 1, \ldots, p$, replace X_j by

$$\frac{X_j - \bar{x}_j}{s_j}$$

and replace Y by

$$\frac{Y - \bar{y}}{s_Y}$$

In the resulting scaled data, the estimate of the intercept is exactly zero, and the estimate of $\hat{\beta}_j$ becomes

$$\hat{\beta}_j(\text{standardized}) = \frac{s_j}{s_Y} \hat{\beta}_j \qquad j = 1, 2, \ldots, p \qquad (7.20)$$

Some investigators compare standardized coefficient estimates for different variables. Under this logic, variables with larger standardized coefficients are more important. Unfortunately, this reasoning is faulty because the scaling depends on the range of values for the variables in the data. For example, if two analysts collect data on the same variables, one collecting data over a small range for the independent variables and the other over a larger range, they may come to completely different conclusions about the relative magnitudes of the standardized coefficients. Also, too much faith is placed in estimated coefficients when the model used is only an approximation to a more complicated relationship.

7.6   Linear transformations and principal components

The invariance of linear regression under location/scale changes is a special case of invariance of regression problems under linear transformation of the independent variables. In a linear transformation, the p independent variables in a model are replaced by up to p (linearly independent) linear combinations of them. We have already seen an example of this in the discussion of the Berkeley Guidance Study in Example 3.1. In that example, the three original variables WT2, WT9, and WT18 were replaced by WT2, WT9 $-$ WT2, and WT18 $-$ WT9, and regression on these three seems to allow a better interpretation of the information available than do the original data.

Another example of linear transformations we have encountered is in the QR factorization in Appendix 2A.3, where a matrix $\mathbf{Q}_1$ with orthogonal columns that are linear combinations of the columns of $\mathbf{X}$ is found. Since $\mathbf{Q}_1$ has orthogonal columns, least squares computations are very simple, and translation from results using $\mathbf{Q}_1$ to the original variables can be done using numerically stable methods.

In general, a (nonsingular) linear transformation of a $n \times p'$ matrix $\mathbf{X}$ is specified by finding a $p' \times p'$ matrix $\mathbf{U}$ (of rank p') so that the transformed variables $\mathbf{Z}$ are given by

$$\mathbf{Z} = \mathbf{X}\mathbf{U} \qquad (7.21)$$

In the guidance study example given above, $\mathbf{U}$ is 4×4 and is given by

$$\mathbf{U} = \begin{bmatrix} 1 & 0 & 0 & 0 \\ 0 & 1 & -1 & 0 \\ 0 & 0 & 1 & -1 \\ 0 & 0 & 0 & 1 \end{bmatrix} \qquad (7.22)$$

By direct multiplication of $\mathbf{Z} = \mathbf{X}\mathbf{U}$, we see that the first column of $\mathbf{Z}$ is the first column of $\mathbf{X}$ (i.e., the column of ones), the second column of $\mathbf{Z}$ is WT2, the third column is WT9 $-$ WT2, and the last column is WT18 $-$ WT9.

If the linear model is $\mathbf{Y} = \mathbf{X}\boldsymbol{\beta} + \mathbf{e}$, and if $\mathbf{U}$ has an inverse, define $\boldsymbol{\alpha} = \mathbf{U}^{-1}\boldsymbol{\beta}$. Then

$$\mathbf{Y} = \mathbf{X}\boldsymbol{\beta} + \mathbf{e}$$
$$= \mathbf{X}(\mathbf{U}\mathbf{U}^{-1})\boldsymbol{\beta} + \mathbf{e}$$
$$= \mathbf{Z}\boldsymbol{\alpha} + \mathbf{e}$$

The least squares estimator of $\boldsymbol{\alpha}$ is $\hat{\boldsymbol{\alpha}} = (\mathbf{Z}^T\mathbf{Z})^{-1}\mathbf{Z}^T\mathbf{Y}$, and the least squares estimator of $\boldsymbol{\beta}$ is $\hat{\boldsymbol{\beta}} = \mathbf{U}\hat{\boldsymbol{\alpha}}$.

Principal components. There is a special transformation of the columns of X where the transformed variables have useful statistical properties. The transformed variables are called principal components. Unlike most other computations in this book, the principal components depend on scaling of the variables, and different scalings lead to different sets of principal components. We must therefore be careful to define all the matrices we use. We assume a model in the deviations from **averages form**

$$\mathcal{Y} = \mathcal{X}\beta^* + e \qquad (7.23)$$

where $\mathcal{X}$ is $n \times p$. To find the principal components, we must find a $p \times p$ orthogonal matrix U such that, if $Z = \mathcal{X}U$ is the resulting matrix of transformed variables, the cross product matrix $Z^T Z = U^T \mathcal{X}^T \mathcal{X} U$ is a diagonal matrix D; that is, we want to find U, such that $UU^T = U^T U = I$ and

$$U^T(\mathcal{X}^T \mathcal{X})U = D$$

Suppose that the diagonal entries of D are $\lambda_1, \lambda_2, \ldots, \lambda_p$ and they are ordered so that $\lambda_1 \geqslant \lambda_2 \geqslant \cdots \geqslant \lambda_p$. The λ_j's are called the eigenvalues of $\mathcal{X}^T \mathcal{X}$. One can show that all the λ_j's are greater than or equal to zero. If $\lambda_p = 0$, then $\mathcal{X}^T \mathcal{X}$ is singular, so examining the smallest eigenvalue is an important check on the near singularity of $\mathcal{X}^T \mathcal{X}$. Computational methods for finding U and the λ_j's are discussed by Stewart (1974), Seber (1977), and Dongarra et al. (1979), the latter source providing FORTRAN code for the calculations. The columns of U are called eigenvectors; the columns of $Z = \mathcal{X}U$ are called principal components.

There are many interpretations and uses of principal components, as in multivariate analysis where the principal components themselves are occasionally used to describe a data set. Also, the eigenvalues, and their relative magnitudes, are useful in other aspects of regression analysis. More complete treatments of the subject are given by Press (1972) and Morrisson (1977).

Example 7.3   Berkeley Guidance Study

For the boys in the Berkeley Guidance Study, using only variables WT2, WT9, and WT18, the matrix of eigenvectors U and the eigenvalues of $\mathcal{X}^T \mathcal{X}$ are

$$U = \begin{bmatrix} 0.0354 & -0.3551 & 0.9341 \\ 0.2754 & -0.8951 & -0.3507 \\ 0.9607 & 0.2697 & 0.0662 \end{bmatrix}$$

$$(\lambda_1 \ \lambda_2 \ \lambda_3) = (3604. \ \ 246.2 \ \ 34.30)$$

The three principal components Z_1, Z_2, and Z_3 are defined by

$$Z_1 = 0.0354 \text{ WT2} + 0.2754 \text{ WT9} + 0.9607 \text{ WT18}$$

$$Z_2 = -.3551 \text{ WT2} - .8951 \text{ WT9} + 0.2697 \text{ WT18}$$

$$Z_3 = 0.9341 \text{ WT2} - 0.3507 \text{ WT9} + 0.0662 \text{ WT18}$$

One can show that the principal components are defined so that, if the model

$$Y = \beta_0 + \beta_1 Z_1 + \beta_2 Z_2 + \beta_3 Z_3 + e$$

were fit with $\text{var}(e) = \sigma^2$, the estimate $\hat{\beta}_1$ would have smaller variance than would the coefficient estimate for any other possible linear combination of the X's and the variance of $\hat{\beta}_1$ will be σ^2/λ_1. Similarly, among all linear combinations of the X's orthogonal to Z_1, the one specified by Z_2 has an estimate $\hat{\beta}_2$ with variance σ^2/λ_2, smaller than for any other linear combination. This is repeated for all other principal components. In the above example, $\text{var}(\hat{\beta}_1) = \sigma^2/3604 = 2.77 \times 10^{-4}\sigma^2$, while $\text{var}(\hat{\beta}_3) = \sigma^2/34.30 = 2.9 \times 10^{-2}\sigma^2$, and much more information is available in the data concerning β_1 than concerning β_3.

If rather than study the principal components based on $\mathcal{X}^T\mathcal{X}$, we compute them from the sample correlation matrix, completely different results are possible. For the example, the matrix of the transformation $\mathbf{U}^*$ and the eigenvalues $\lambda_1^*, \lambda_2^*, \lambda_3^*$ are

$$\mathbf{U}^* = \begin{bmatrix} 0.4925 & -0.7809 & -0.3843 \\ 0.6648 & 0.0525 & 0.7452 \\ 0.5618 & 0.6224 & -0.5450 \end{bmatrix}$$

$$(\lambda_1^* \;\; \lambda_2^* \;\; \lambda_3^*) = (2.028 \;\; 0.7890 \;\; 0.1829)$$

Approximately, then, by rounding of coefficients, the three principal components are

$$Z_1 = 0.4925 \text{ WT2} + 0.6648 \text{ WT9} + 0.5618 \text{ WT18}$$

$$\cong \tfrac{1}{2}(\text{WT2} + \text{WT9} + \text{WT18})$$

$$= \text{measure of average weight}$$

$$Z_2 = -0.7809 \text{ WT2} + 0.0525 \text{ WT9} + 0.6224 \text{ WT18}$$

$$\cong 0.7(-\text{WT2} + \text{WT18})$$

$$= \text{linear weight gain from age 2 to age 18}$$

$$Z_3 = -0.3843 \text{ WT2} + 0.7452 \text{ WT9} - 0.5450 \text{ WT18}$$

$$\cong 0.4(-\text{WT2} + 2\text{WT9} - \text{WT18})$$

$$= \text{quadratic weight gain}$$

Thus in the correlation form, the principal components have very simple explanations, while simple explanations are not available for the computations based on $\mathcal{X}^T\mathcal{X}$. Those computations suggest that the first principal component—the combination of X's for which a coefficient is most precisely estimated—is approximately equal to $0.3[\text{WT9} + 2(\text{WT18})]$. The importance of WT18 in these results is due to the fact that the sample variance of WT18 is largest among the variances of WT2, WT9, and WT18, and larger variances give WT18 more weight in the principal component. In the correlation form, differences between variances have been removed, and all three variables are used in a seemingly useful way.

Principal components of many variables. Principal components can be found for any set of variables, such as the predictors of SOMA in the Berkeley Guidance Study. However, the resulting variables will be linear combinations of dissimilar variables—heights, weights, and strengths—and the new variables may have no meaning. An alternative approach that would apply to the guidance study would be to find the principal components for the height variables separately from the others, the weights separately from the others, and so on, using these transformed regressors in further analyses.

Problems

7.1 As an alternative to fitting the model

$$Y = \beta_0 + \beta_1 X_1 + \beta_2 X_2 + \beta_{12} X_1 X_2 + \text{error}$$

consider fitting the model

$$Y = \beta_0 + \beta_1 X_1 + \beta_2 X_2 + \beta_3 (X_1 - X_2) + \text{error}$$

Discuss the differences between these two models. Under what circumstances is the second one appropriate? When is the first one better? (Hint: What happens to the response in each model if X_1 is changed to $X_1 + \delta_1$ and/or X_2 is changed to $X_2 + \delta_2$?)

7.2 Compare the regression lines for Forbes' data (Example 1.1) and Hooker's data (Problem 1.2)

7.3 In the Berkeley Guidance Study data, Problem 2.1, consider modeling HT18 by the data collected at ages 2 and 9. Fit the same model for boys and girls and perform appropriate tests to compare the fitted regression planes.

7.4   For the girls in the Berkeley Guidance Study data, find the principal components for the three weight variables, paralleling the computations for boys given in the text. Comment on the meaning, if any, of the principal components.

7.5   For the physics data (Problem 4.2), formally compare the regression lines suggested by Problem 4.2.1.

7.6   For the apple shoot data, Example 4.2, perform a complete data analysis.

8

MODEL BUILDING II: COLLINEARITY AND VARIABLE SELECTION

Suppose that k independent variables $X_1, X_2, \ldots, X_k$ are available to model a response Y. Some of the X's may be exact functions of others such as $X_5 = X_4^2$, $X_6 = X_1 X_2$. Under the guiding principle that fewer is better—an empirical model using a few predictors is more useful than one using more predictors—finding a subset of the X's to model Y may be desirable. Besides the obvious fact that understanding a smaller model is usually easier than understanding a larger one, we shall see that if some of the predictors are related to each other, models based on subsets may give more precise results than will models based on more variables.

Variable selection is a desirable part of many, but not all, regression analyses. The techniques presented in this chapter help to identify several subsets of variables that are equivalent or nearly equivalent in their ability to model a response. Choice between these final few models, if a choice is possible, is made on the basis of case analyses, sensibility of coefficients, interpretation of the results, and similar considerations. No automatic procedure for finding a single best model is proposed.

8.1 Collinearity

Two variables X_1 and X_2 are exactly collinear in a data set if there is a linear equation such as $X_2 = \alpha_0 + \alpha_1 X_1$ for some constants α_0 and α_1 that is true for all cases in the data. For example, suppose that X_1 and X_2

174

represent the amounts of two different chemicals used in an experiment. If, in the experiment, the sum of X_1 and X_2 is fixed, perhaps, $X_1 + X_2 = 50$ ml, then for any choice of $X_1 \leqslant 50$ ml, $X_2 = 50 - X_1$, and X_1 and X_2 are exactly collinear.

Approximate collinearity is obtained if the equation $X_2 = \alpha_0 + \alpha_1 X_1$ holds approximately for the observed data. A measure of the degree of collinearity between X_1 and X_2 is the square of the sample correlation r_{12}^2 between the two variables. Exact collinearity corresponds to $r_{12}^2 = 1, r_{12}^2 = 0$ corresponds to noncollinearity, and the closer r_{12}^2 is to 1, the stronger the approximate collinearity. In the above example, if the experiment were done so that $X_1 + X_2$ is approximately 50 ml, then X_1 and X_2 would exhibit approximate collinearity. Usually, the adjective "approximate" is dropped, and we would simply say that X_1 and X_2 are collinear if r_{12}^2 is large.

Now, suppose that we want to fit a regression model of a response Y on X_1 and X_2 using the model

$$y_i = \beta_0 + \beta_1 x_{i1} + \beta_2 x_{i2} + e_i \qquad i = 1, 2, \ldots, n$$

with $\text{var}(e_i) = \sigma^2$, and all the e_i are uncorrelated. We now ask how the properties of the fitted regression equation will change as r_{12}^2 increases from 0 to 1. Suppose that estimation of the parameter β_1 is of interest, although most other quantities, such as fitted values and predictions, will lead to equivalent results. For simplicity, assume that X_1 and X_2 have been scaled so that $\sum(x_{i1} - \bar{x}_1)^2 = \sum(x_{i2} - \bar{x}_2)^2 = 1$.

As long as $r_{12}^2 < 1$, the least squares estimator $\hat{\beta}_1$ of β_1, found from equation (2.15), is unbiased. It is left as an exercise to show that the variance of $\hat{\beta}_1$ for the full model, say, $\text{var}(\hat{\beta}_1 \mid \text{full model})$, is

$$\text{var}(\hat{\beta}_1 \mid \text{full model}) = \sigma^2 \frac{\sum(x_{i1} - \bar{x}_1)^2}{1 - r_{12}^2} = \frac{\sigma^2}{1 - r_{12}^2} \qquad (8.1)$$

When X_1 and X_2 are uncorrelated, $r_{12}^2 = 0$, $\text{var}(\hat{\beta}_1 \mid \text{full model})$ has its minimum value. As r_{12}^2 approaches 1, $\text{var}(\hat{\beta}_1)$ will get to be arbitrarily large, and $\hat{\beta}_1$ will be unacceptably variable, as will predictions and other functions of $\hat{\beta}_1$.

The case of $r_{12}^2 = 1$ must be treated separately because if one variable can be obtained exactly as a linear function of another variable, then we cannot find a unique estimator $\hat{\beta}_1$ as the matrix $\mathbf{X}^T\mathbf{X}$ in (2.15) is not invertable. To obtain a unique least squares estimator, one or the other of the two variables must be removed from the model, as both independent variables measure exactly the same thing.

Now, compare fitting a subset model $y_i = \beta_0 + \beta_1 x_{i1} + e_i, i = 1, \ldots, n$, obtained from the full model by deleting X_2. Generally, if the full model is correct (in the sense that it is unbiased) then the estimate of β_1 from the subset model is biased and, using the scaling for the X's given previously,

$$E(\hat{\beta}_1 \,|\, \text{subset model}) = \beta_1 + r_{12}\beta_2 \qquad (8.2)$$

The bias in estimating β_1 from the subset model is $\beta_1 - (\beta_1 + r_{12}\beta_2) = - r_{12}\beta_2$, and the variance of $\hat{\beta}_1$ from the subset model can be shown to be

$$\text{var}(\hat{\beta}_1 \,|\, \text{subset model}) = \sigma^2 \qquad (8.3)$$

which does not depend upon r_{12}. To compare estimation of β_1 from the full and subset models, we must compute the mean square error of $\hat{\beta}_1$ given the subset model, mse($\hat{\beta}\,|\,$ subset model), where the mse is defined by

$$\text{mse}(\hat{\beta}_1 \,|\, \text{subset model}) = \text{var}(\hat{\beta}_1 \,|\, \text{subset model}) + (\text{bias})^2$$

Then

$$\text{mse}(\hat{\beta}_1 \,|\, \text{subset model}) = \sigma^2 + (r_{12}\beta_2)^2 \qquad (8.4)$$

Comparing (8.1) and (8.4), we see that the subset model will estimate β_1 more precisely (i.e., mse($\hat{\beta}_1\,|\,$ subset model) $<$ var($\hat{\beta}_1\,|\,$ full model)) whenever

$$\frac{|\beta_2|}{\sigma} < \frac{1}{\sqrt{1 - r_{12}^2}} \qquad (8.5)$$

The result (8.5) is shown graphically in Figure 8.1. If r_{12}^2 is near 1, the subset model is almost always better than the full model, while for any value of r_{12}^2, the subset model will be preferred if $|\beta_2| < \sigma$. Thus when collinearity is observed in the data more precise estimation of β_1 can generally be obtained from the subset model than from the full model, unless the coefficient for the deleted variable is very large. If the β's and σ^2 were known, deletion of variables with small $|\beta_j|/\sigma$ would be desirable. Since these quantities are generally unknown, selection techniques should have the property of deleting variables with $|\beta_j|/\sigma$ probably small.

Collinearities between two variables can be diagnosed by examining the sample correlation matrix between the potential predictor variables. However, only dependencies between pairs of variables can be diagnosed by this examination. Also, collinearity is a function of the sample correlations, which, depending on the methods of collecting data, may be quite different from the correlations between the quantities in some hypothetical population. Thus two variables that are thought to be completely separate and unrelated in theory may appear to be highly collinear in a practical problem, and inclusion of both in a regression analysis can lead to poor estimates.

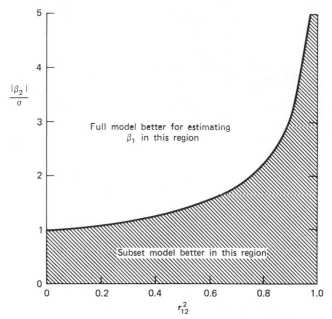

Figure 8.1 Comparing a one-predictor model to a two-predictor model.

More general collinearity. Collinearities between several variables, sometimes called multicollinearities, can occur if a nearly exact linear relationship exists between several independent variables. Suppose that c is a vector of length 1 ($c^T c = 1$). The vector c is called a direction vector. We will say that collinearity exists between columns of a matrix X if, for some c, the product Xc is nearly 0 (Silvey, 1969). If x_i^T is the ith row of X, then collinearity means each $x_i^T c$ is nearly equal to zero, and therefore, if the values of all but one of the elements of x_i^T are known, and c is known, the value of the remaining element of x_i^T can be found almost exactly as a linear combination of the other elements of x_i. Equivalently, at least one column of X can be very closely approximated by a linear combination of the other columns of X.

To measure collinearity, it is convenient to consider the squared length of Xc, or $(Xc)^T Xc = c^T (X^T X)c$, and we will say that collinearity exists if $c^T (X^T X)c$ is sufficiently close to zero for some c. One can show that for any choice of c, $c^T (X^T X)c$ is greater than or equal to the smallest eigenvalue of $X^T X$ (equality occurs if c is taken to be the eigenvector corresponding to the smallest eigenvalue). Thus collinearity will be diagnosed if the smallest eigenvalue is sufficiently small relative to the other eigenvalues. Judging the size of the smallest eigenvalue is difficult since it depends on the scaling of the columns of $\mathbf{X}$. As pointed out in Section 7.6, the eigenvalues of $\mathbf{X}^T\mathbf{X}$, $\mathcal{X}^T\mathcal{X}$, and the sample correlation matrix may all be substantially different. Typically, the eigenvalues of $\mathcal{X}^T\mathcal{X}$ will be used in preference to those of $\mathbf{X}^T\mathbf{X}$. If the units of measurement of the columns of $\mathbf{X}$ are arbitrary, then the eigenvalues of the correlation matrix are usually computed. If $\mathcal{X}^T\mathcal{X}$ is used, columns of $\mathcal{X}$ with large sample variances will be more important in determining the eigenvalues than will columns of $\mathcal{X}$ that have small sample variance, and variations in the sample variances will be reflected as strongly in the eigenvalues as will collinearity. Basing computations on the correlation matrix has the effect of eliminating the effect of differences in variances.

One popular measure of collinearity is called the *condition number*, κ, defined by

$$\kappa = (\text{largest eigenvalue}/\text{smallest eigenvalue})^{1/2} \qquad (8.6)$$

Clearly, $\kappa \geqslant 1$, with large values suggesting collinearity. The condition number arises as a natural summary statistic in the analysis of numerical properties of least squares algorithms (Stewart, 1974). Unfortunately, because of the problem of lack of invariance of κ under changes in scale of the X's, interpretation of the condition number is difficult.

Before turning to an example, it is useful to distinguish further between numerical collinearity and statistical collinearity. A data set that exhibits strong enough collinearity to make accurate computations impossible will be called ill-conditioned. Precise definition of ill-conditioning depends on the computer and the algorithms used. Computers with 10 digits of accuracy, for example, will require a higher degree of collinearity for ill-conditioning than will a computer with 7 digits of accuracy, and, similarly, highly accurate algorithms may find least square estimates in problems that less accurate algorithms could not handle. When a problem is ill-conditioned, reliable least squares computations may not be possible.

The relative sensitivity coefficients discussed in Section 3.3 are useful for diagnosing ill-conditioning. Setting each of the σ_j^2's in (3.14) to represent round-off errors in the last significant digit, if each g_{jk} is small, say less than 10^{-1}, collinearity is not numerically important.

Collinearity that is statistically important may be much less severe than the collinearity required for ill-conditioning. Thus even if accurate computations are done, the variances of some statistics, such as parameter estimates or predictions, may be so large as to make them practically useless. On the other hand, other aspects of a problem—estimates of other parameters or predictions at other points—may be perfectly reliable.

Many treatments have been suggested for data that exhibit collinearity. The method described in this chapter is to delete some of the variables from a fitted model (other methods are outlined in Chapter 11). It is a pleasing fact that, if variables are deleted from a model, the condition number of the subset model is always less than the condition number of a larger model. Thus the treatment proposed here does improve the observed symptom.

Example 8.1   Highway data

This data set, given in Table 8.1 and taken from an unpublished master's paper in civil engineering by Carl Hoffstedt, relates the automobile accident rate, in accidents per million vehicle miles (Y) to 13 potential independent variables. The data include 39 sections of large highways in the state of Minnesota in 1973. The variables in the order given in Table 8.1, are:

Y = RATE = 1973 accident rate per million vehicle miles.

$X1$ =   LEN = length of the segment in miles.

$X2$ =   ADT = average daily traffic count in thousands (estimated).

$X3$ = TRKS = truck volume as a percent of the total volume.

$X4$ =   SLIM = speed limit (in 1973, before the 55 mph limits).

$X5$ = LWID = lane width in feet.

$X6$ = SHLD = width in feet of outer shoulder on the roadway.

$X7$ =   ITG = number of freeway-type interchanges per mile in the segment.

$X8$ =   SIGS = number of signalized interchanges per mile in the segment.

$X9$ = ACPT = number of access points per mile in the segment.

$X10$ = LANE = total number of lanes of traffic in both directions.

$X11$ =   FAI = 1 if Federal aid interstate highway, 0 otherwise.

$X12$ =   PA = 1 if principal arterial highway, 0 otherwise.

$X13$ =   MA = 1 if minor arterial highway, 0 otherwise.

Table 8.1 Highway data

	Y	X_1	X_2	X_3	X_4	X_5	X_6	X_7	X_8	X_9	X_{10}	X_{11}	X_{12}	X_{13}
1	4.58	4.99	69	8	55	12	10	1.20	0	4.60	8	1	0	0
2	2.86	16.11	73	8	60	12	10	1.43	0	4.40	4	1	0	0
3	3.02	9.75	49	10	60	12	10	1.54	0	4.70	4	1	0	0
4	2.29	10.65	61	13	65	12	10	0.94	0	3.80	6	1	0	0
5	1.61	20.01	28	12	70	12	10	0.65	0	2.20	4	1	0	0
6	6.87	5.97	30	6	55	12	10	0.34	1.84	24.80	4	0	1	0
7	3.85	8.57	46	8	55	12	8	0.47	0.70	11.00	4	0	1	0
8	6.12	5.24	25	9	55	12	10	0.38	0.38	18.50	4	0	1	0
9	3.29	15.79	43	12	50	12	4	0.95	1.39	7.50	4	0	1	0
10	5.88	8.26	23	7	50	12	5	0.12	1.21	8.20	4	0	1	0
11	4.20	7.03	23	6	60	12	10	0.29	1.85	5.40	4	0	1	0
12	4.61	13.28	20	9	50	12	2	0.15	1.21	11.20	4	0	1	0
13	4.80	5.40	18	14	50	12	8	0	0.56	15.20	2	0	1	0
14	3.85	2.96	21	8	60	12	10	0.34	0	5.40	4	0	1	0
15	2.69	11.75	27	7	55	12	10	0.26	0.60	7.90	4	0	1	0
16	1.99	8.86	22	9	60	12	10	0.68	0	3.20	4	0	1	0
17	2.01	9.78	19	9	60	12	10	0.20	0.10	11.00	4	0	1	0
18	4.22	5.49	9	11	50	12	6	0.18	0.18	8.90	2	0	1	0
19	2.76	8.63	12	8	55	13	6	0.14	0	12.40	2	0	1	0
20	2.55	20.31	12	7	60	12	10	0.05	0.99	7.80	4	0	1	0
21	1.89	40.09	15	13	55	12	8	0.05	0.12	9.60	4	0	1	0
22	2.34	11.81	8	8	60	12	10	0	0	4.30	2	0	1	0
23	2.83	11.39	5	9	50	12	8	0	0.09	11.10	2	0	1	0
24	1.81	22.00	5	15	60	12	7	0	0	6.80	2	0	1	0
25	9.23	3.58	23	6	40	12	2	0.56	2.51	53.00	4	0	0	1
26	8.60	3.23	13	6	45	12	2	0.31	0.93	17.30	2	0	0	1
27	8.21	7.73	7	8	55	12	8	0.13	0.52	27.30	2	0	0	1
28	2.93	14.41	10	10	55	12	6	0	0.07	18.00	2	0	0	1
29	7.48	11.54	12	7	45	12	3	0.09	0.09	30.20	2	0	0	1
30	2.57	11.10	9	8	60	12	7	0	0	10.30	2	0	0	1
31	5.77	22.09	4	8	45	11	3	0	0.14	18.20	2	0	0	1
32	2.90	9.39	5	10	55	13	1	0	0	12.30	2	0	0	1
33	2.97	19.49	4	13	55	12	4	0	0	7.10	2	0	0	1
34	1.84	21.01	5	12	55	10	8	0	0.10	14.00	2	0	0	1
35	3.78	27.16	2	10	55	12	3	0.04	0.04	11.30	2	0	0	1
36	2.76	14.03	3	8	50	12	4	0.07	0	16.30	2	0	0	1
37	4.27	20.63	1	11	55	11	4	0	0	9.60	2	0	0	1
38	3.05	20.06	3	11	60	12	8	0	0	9.00	2	0	0	0
39	4.12	12.91	1	10	55	12	3	0	0	10.40	2	0	0	0

Table 8.2 Summary statistics

(a) Means and variances

Variable	N	Mean	Variance	Standard Deviation	Minimum	Maximum
RATE	39	3.933	3.944	1.986	1.610	9.230
LEN	39	12.88	57.91	7.610	2.960	40.09
ADT	39	19.62	346.4	18.61	1.000	73.00
TRKS	39	9.333	5.544	2.355	6.000	15.00
SLIM	39	55.00	34.21	5.849	40.00	70.00
LWID	39	11.95	0.2078	0.4559	10.00	13.00
SHLD	39	6.972	9.220	3.036	1.000	10.00
ITG	39	0.2964	0.1691	0.4112	0	1.540
SIGS	39	0.4005	0.4012	0.6334	0	2.510
ACPT	39	12.16	86.83	9.318	2.200	53.00
LANE	39	3.128	1.852	1.361	2.000	8.000
FAI	39	0.1282	0.1147	0.3387	0	1.000
PA	39	0.4872	0.2564	0.5064	0	1.000
MA	39	0.3333	0.2281	0.4776	0	1.000

(b) Sample correlation matrix

	RATE	LEN	ADT	TRKS	SLIM	LWID	SHLD	ITG	SIGS	ACPT	LANE	FAI	PA	MA
RATE	1.00													
LEN	−0.47	1.00												
ADT	−0.03	−0.27	1.00											
TRKS	−0.51	0.50	−0.10	1.00										
SLIM	−0.68	0.19	0.24	0.30	1.00									
LWID	−0.01	−0.31	0.13	−0.15	−0.10	1.00								
SHLD	−0.39	−0.10	0.46	0.00	0.69	−0.04	1.00							
ITG	−0.02	−0.25	0.90	−0.07	0.24	0.10	0.38	1.00						
SIGS	0.56	−0.32	0.15	−0.45	−0.41	0.04	−0.13	−0.07	1.00					
ACPT	0.75	−0.24	−0.22	−0.36	−0.68	0.04	−0.43	−0.20	0.50	1.00				
LANE	−0.03	−0.20	0.82	−0.15	−0.26	−0.04	0.48	0.70	0.25	−0.21	1.00			
FAI	−0.20	−0.03	0.76	0.14	0.46	0.23	0.40	0.81	−0.25	−0.34	0.59	1.00		
PA	−0.16	−0.15	−0.03	−0.05	0.04	0.37	−0.13	0.30	0.30	−0.23	0.17	−0.37	1.00	
MA	−0.34	0.13	−0.46	0.10	−0.42	−0.28	−0.62	0.36	−0.07	0.51	−0.51	−0.27	−0.69	1.00

Table 8.3 Regression for full model

Variable	Estimate	Standard Error	t-Value
Intercept	13.7	6.87	1.99
LEN	− 0.065	0.033	− 1.94
ADT	− 0.040	0.034	− 0.12
TRKS	− 0.100	0.011	− 0.87
SLIM	− 0.124	0.082	− 1.52
LWID	− 0.134	0.598	− 0.22
SHLD	0.014	0.162	0.09
ITG	− 0.475	1.28	− 0.37
SIGS	0.713	0.525	1.36
ACPT	0.067	0.043	1.56
LANE	0.027	0.283	0.09
FAI	0.543	1.72	0.31
PA	− 1.01	1.11	− 0.91
MA	− 0.548	0.976	− 0.56

$\hat{\sigma}^2 = 1.44$, d.f. $= 25$, $R^2 = 0.76$, $RSS = 35.89367$.

Two of the highway segments, numbers 38 and 39, were neither interstate highways, principal arterial, nor major arterial highways, but were classified as major collectors (MC) and are coded by $FAI = PA = MA = 0$. A separate variable was not used for MC because the resulting data matrix would give an exact collinearity and one of the dummy variables would have to be deleted to obtain estimates.

To find the condition number of the sample correlation matrix of the X's, we first find the eigenvalues to be

$$4.50, 2.70, 2.03, 1.07, 0.84, 0.52, 0.44$$
$$0.31, 0.24, 0.16, 0.078, 0.060, 0.051$$

from which we find the condition number $\kappa = (4.50/.051)^{1/2} = 9.4$, indicating modest collinearity in this data set. However, eight or nine of the eigenvalues are relatively small, and we might expect that deletion of about that many variables will result in very little loss of information.

Another simple diagnostic for statistically important collinearity is examination of the t-values for each of the coefficient estimates in the full model, as shown in Table 8.3 (Table 8.2 gives the mean standard deviations and correlations of all the variables). This table exhibits the curious property that, while $R^2 = 0.76$, none of the t-values exceeds 2 in absolute value. Recall that the t-values measure the effect of a variable adjusted for all the others. Each of these small values indicates that the unique portion of Y explained by any one X but not by the others is very small. This also indicates the need for deleting variables from the model.

8.2 Assumptions and notation

On each of n cases, we observe k independent variables $X_1, \ldots, X_k$, and a dependent variable Y. In earlier chapters, the number of variables is p; in this chapter we reserve p for the number of variables in a selected subset, and let $k' = k + 1$ if the model includes a constant term and $k' = k$ if regression is through the origin.

The full model using all the independent variables is written in matrix terms as

$$Y = X\beta + e \quad \begin{cases} X: n \times k' \\ \beta: k' \times 1 \\ e: n \times 1 \quad (\text{cov}(e) = \sigma^2 I) \end{cases} \quad (8.7)$$

If $n < k'$, we have more variables than cases, and we will not be able to find a unique estimate of β; however, model (8.7) still can provide a description of a relationship between Y and the X's. Now, a subset model can be specified by partitioning X into two matrices X_1 and X_2, so that X_1 is $n \times p$ with $p \leqslant n$ (and rank$(X_1) = p$) and X_2 is $n \times (k' - p)$. X_1 consists of the variables in the subset model, and X_2 consists of the variables not included in the subset model. Usually, X_1 will include the constant (i.e., the column of 1's), if any, although the constant could be treated like any other variable in the regression problem.

Corresponding to the partition of X into X_1 and X_2 we partition β into β_1 and β_2 where β_1 is $p \times 1$ and β_2 is $(k' - p) \times 1$. The full model (8.7) can be rewritten as

$$Y = X_1 \beta_1 + X_2 \beta_2 + e \quad (8.8)$$

Next, a (specific) subset model is obtained by deleting the term $X_2 \beta_2$ to give

$$Y = X_1 \beta_1 + e^* \quad (8.9)$$

The two models (8.8) and (8.9) are identical only if $\beta_2 = 0$. In general, these two models are quite different, as the estimate of the parameter β_1 will usually be different in each of the models, and the interpretation of the parameter depends on which model is used. Also, by restricting ourselves to subsets with $p \leqslant n$, we can always obtain unique estimates for the parameters in (8.9) even though unique estimates do not exist for (8.8) if $k' > n$.

To proceed, assume that all relevant variables that could be included, plus perhaps a few irrelevant ones, are included in the set of k possible independent variables. This means that estimates of fitted values from the full model (8.8) will be unbiased, although, if some irrelevant variables are included in the study, they will not be as precise as possible. In practical terms, this assumption can be weakened by limiting attention to study of the relationship for cases like those in the data. As long as extrapolation outside the observed range of data is not of interest, and allowing data to be transformed as necessary, the full model will generally give fitted values that are at least nearly unbiased over the limited range.

In practice, correct scales for variables are often unknown, and an analyst must decide both on scaling and on model selection. As a general approach, choice of transformation and diagnosing other problems through the use of case statistics can be done before subset selection. However, repeating case analysis after subset selection will always be prudent.

In the subset model, estimate $\boldsymbol{\beta}_1$ by $\hat{\boldsymbol{\beta}}_1 = (\mathbf{X}_1^T\mathbf{X}_1)^{-1}\mathbf{X}_1^T\mathbf{Y}$, as if least squares were being used and no other X's were observed.

8.3   Selecting subsets on substantive grounds

The single most important tool in selecting a subset of variables for use as a model is the analyst's knowledge of the substantive area under study and of each of the variables, including expected sign and magnitude of the coefficient.

In the highway accident data, there are $k = 13$ potential predictors, so there are $2^{13} = 8192$ possible subset models (including the full model and the model containing only the intercept). However, the 13 independent variables can be divided into several types. First of all, variables 11, 12, and 13, namely, FAI, PA, and MA, are simply indicator variables that taken together indicate the type of highway. It may not be reasonable to include one of these variables while omitting the other two (this is partly because the fourth type of highway, major collectors, is indicated by $FAI = PA = MA = 0$). Thus we may consider models with either all three included or all three excluded. Possibly, these may have special importance, since the type of roadway is defined by the source of financial support that the Highway Department uses to maintain the roads. We may even think of this problem as being an analysis of covariance, where we explore the possibility of a difference between highway types adjusted by the other independent variables.

Also, the variable LEN should be treated differently from the others, since its inclusion in the prediction equation may be required by the way highway segments are defined. Suppose that highways consist of "safe stretches" and "bad spots," and that most accidents occur at the bad spots. If we were to lengthen a highway segment in our study by one mile, it is unlikely that we would add another bad spot to the section, assuming bad spots are rare. However, as a result of lengthening the roadway in the study, the computed response, accidents per million vehicle miles on the section of roadway, would have a lower value (since the number of miles driven would go up, and the number of accidents would stay about the same). Thus the response and LEN should be negatively correlated, and we should consider only models that include LEN.

The number of possible subset models for the highway data is now reduced from 8192 to 512 with LEN and the type indicators, and 512 with LEN but without the type indicators, which is a more manageable problem, although still quite large.

Examination of the signs of the estimated coefficients in the full model may also lead to decisions concerning which variables should or should not be used. In Table 8.3, for example, the fitted coefficient estimates lead to the following statements: Lower accident rates occur with more truck traffic, higher speed limits, wider lanes, more interchanges, fewer signals, and so on. Many of these statements seem against intuition, and may lead to deletion of variables.

Redefining variables. Another important method of reducing the number of independent variables is to define new variables that are combinations of the old variables. For example, in some studies, the two variables height and weight, which are often highly correlated, may be replaced by a combination of them to give a single height-weight index, or two IQ tests as independent variables can be replaced by their average.

8.4 Finding subsets

In many problems, a point is reached at which it is necessary to actually use the data to find subsets of the variables. There are two basic problems. First, a criterion is required for deciding if one subset is better than another. The second problem is computational, since the number of possible subsets that must be considered can be enormous. We first discuss the problem of choosing a criterion of comparison. Most analysts have traditionally used ad hoc criteria based on summary statistics, such as R^2. After a brief review of these, we consider more systematic criteria based on mean square error of prediction. Computational considerations are left for the next section.

Ad hoc criteria for model selection. For purposes of an example, consider a fixed subset model for the highway data consisting of LEN, SLIM, SIGS, ACPT (and the intercept) as being the subset of interest. No claim is made that this is a particularly good subset. In fact, the subset was chosen by including all variables with the t-values in Table 8.3 that are greater than 1 in absolute value (this is not a recommended technique). The summary of the regression for the subset model is given in Table 8.4. Generally, ad hoc criteria are based on a few of the summary statistics for the subset model and for the full model. Let RSS_p and $RSS_{k'}$ be, respectively, the residual sum of squares for the subset and the full models, and let TSS be the total sum of squares. For all of the statistics covered, the subscript p refers to the number of parameters in the subset under consideration; in the example, $p = 4 + 1 = 5$.

Table 8.4   A subset model

Variable	Estimate	Standard Error	t-Value
Intercept	8.81	2.60	3.38
LEN	− 0.069	0.025	− 2.72
SLIM	− 0.096	0.043	− 2.26
SIGS	0.485	0.342	1.42
ACPT	0.089	0.028	3.17

$\hat{\sigma}^2 = 1.25$, d.f.= 34, $R^2 = 0.72$, $RSS = 42.3333$.

The F-test. The subset model can be compared to the full model by use of the F-test that $\beta_2 = 0$. The F-test can be used to compare any two models as long as all of the parameters in the smaller model are also included in the larger model; if this is so, we say that the smaller model is a submodel of the larger model. Clearly, a subset model is a submodel of the full model, so we can do the F-test.

A convenient form for the F-test is given by

$$F_p = \frac{(RSS_p \quad RSS_{k'})/(k' - p)}{(RSS_{k'})/(n - k')}$$

(8.10)

For the highway example, the statistic F_p is given by

$$F_p = \frac{(42.33 - 35.89)/9}{35.89/25} = 0.50$$

We could then have a rule to prefer a full model (or, in general the larger model) to the subset model if F_p is sufficiently large. One reasonable rule would be to prefer the full model if $F_p > F^*$, where F^* is the $\alpha \times 100\%$ point of the $F(k' - p, n - k')$ distribution; the choice of $\alpha = 0.05$ is not unusual. This rule assumes that the error vector $\mathbf{e}$ is normally distributed. In the example, since $F_p = 0.50$, we would almost certainly prefer the subset model to the full model based on this criterion.

If more than two models are to be considered, the F_p-test cannot be directly applied, since we can only compare models where one is a submodel of the other. This excludes, for example, comparison of two models of the same size p. Also, F_p measures the overall closeness of $\boldsymbol{\beta}_2$ to $\mathbf{0}$; if one of the components of $\boldsymbol{\beta}_2$ is far from zero, F_p may still be small.

Coefficient of determination or multiple correlation coefficient. The square of the multiple correlation coefficient (R^2) is used as a criterion for comparing models. Recall that R^2 is the square of the correlation between the dependent variable Y and the best single linear combination of the independent variables. A computing formula for R^2 in a p-parameter model is

$$R_p^2 = 1 - \frac{RSS_p}{TSS} \tag{8.11}$$

R_p^2, which is between 0 and 1, is the proportion of variability in Y explained by regression on the X's, the greater the value of R_p^2, the more variability is explained. Unfortunately, R_p^2 provides an inadequate criterion for subset selection since, whenever comparing a subset model to a larger model including the subset, the larger model will always have a value of R^2 as large, or larger, than R^2 for the subset model. Thus the full model will always have the largest possible value of R^2. However, for fixed p, R^2 can be used to compare different models, large value of R_p^2 indicating preferred models.

For the highway data, $R_{k'}^2 = 0.7605$, while for the subset, $R_p^2 = 0.7176$. By eliminating 9 of the independent variables, we have only introduced an additional 4% of unexplained variability; usually, such a reduction would be judged to be small compared to the savings gained by deleting so many variables.

Adjusted R^2. Since R^2 itself is nondecreasing in the number of parameters p, one may choose to use an adjusted version of R^2 defined by

$$\overline{R}_p^2 = 1 - \left(\frac{n-1}{n-p} \right)(1 - R_p^2) \tag{8.12}$$

(If the constant is not included in the subset model, then the $n - 1$ in (8.12) should be replaced by n.) Good models will have large values of $\overline{R}_p^2$. For the highway data, $\overline{R}_{k'}^2 = 0.636$ for the full model, and $\overline{R}_p^2 = 0.684$ for the subset model, suggesting that the subset model is somewhat better on this criterion. Note that $\overline{R}^2$ can be negative.

Residual mean square. A related criterion is to minimize the residual mean square $\hat{\sigma}_p^2$ for the models under consideration, where

$$\hat{\sigma}_p^2 = RSS_p/(n - p) \tag{8.13}$$

This criterion need not select the full model, since both RSS_p and $n - p$ will decrease as p increases, and (8.13) may decrease or increase. In most cases, however, this criterion will favor large subsets over smaller ones. For the highway data, $\hat{\sigma}_{k'}^2 = 1.436$ for the full model, while for the subset model, $\hat{\sigma}_p^2 = 1.245$, so that the subset model is better than the full model on this criterion.

We can relate this criterion to the F_p given by (8.10) since $\hat{\sigma}_p^2 < \hat{\sigma}_{k'}^2$ if and only if $F_p < 1$. Thus the residual mean square criterion is equivalent to the F_p criterion (for comparing a submodel to a larger model) with the critical value $F^* = 1$ rather than a percentage point of an F-distribution. In the example, recall that $F_p = 0.50$, which is smaller than 1. In fact, simple relationships between $\overline{R}_p^2$, R_p^2, $\hat{\sigma}_p^2$, and F_p are easily found.

Mallows' C_p statistic. Suppose that the goal of regression is estimation of fitted values (or, equivalently, prediction). The precision of estimation of fitted values depends on two distinct sources: variance of the estimates and bias of the estimates. Suppose we let $\hat{y}_i$ be the fitted value for case i obtained from the subset model (8.9). We prefer subset models to make the mean square errors, $\text{mse}(\hat{y}_i)$, $i = 1, 2, \ldots, n$, as small as possible. An important criterion of this type will make

$$E(J_p) = \frac{1}{\sigma^2} \sum_{i=1}^{n} \text{mse}(\hat{y}_i) = \frac{1}{\sigma^2} \sum_{i=1}^{n} \left(\text{var}(\hat{y}_i) + [E(\hat{y}_i) - E(y_i)]^2 \right) \tag{8.14}$$

as small as possible. Note that use of J_p places special emphasis on the observed data. If, for example, one point is replicated on several cases, then the mse at that point will receive a large weight in J_p. This is reasonable if the cases are a sample from a population for which predictions of future values are required, or if interest genuinely centers on the n cases in the data. In other circumstances, a different function of the $mse(\hat{y}_i)$ might be preferable (Mallows, 1973).

Mallows' C_p is an estimate of J_p obtained by using observed data to estimate σ^2 and $mse(\hat{y}_i)$. From Appendix 8A.1, the estimate of J_p is

$$C_p = \frac{RSS_p}{\hat{\sigma}^2} + 2p - n$$

$$= \frac{RSS_{k'} - RSS_p}{\hat{\sigma}^2} + p - (k' - p) \qquad (8.15)$$

where $\hat{\sigma}^2$ is usually computed from the fit of the full model. Mallows suggests that good models will have negative or small $C_p - p$.

For the highway data, $C_p = (42.33/1.436) + 10 - 39 = 0.48$ and $C_p - p = 0.48 - 5 = -4.52$, indicating that this particular subset model could be used without any estimated loss in total mse. This subset model would be preferred to the full model, and is a competitor when compared to all other models.

General mean square error criteria. Other mean square error criteria might prefer subset models that give small mse for the fitted value or prediction at a specific point that need not be one of the rows of X, or, for all points in some region. The following results are for general mse criteria.

Result 1. For any subset model, there always exist some points where the subset model has lower mse than the full model. Thus the full model is never better for all points.

It is easy to find points where the subset model is better by taking all of the independent variables not in the model to be equal to their sample averages. Then, the deleted variables carry no information, so it should be no surprise that the subset model is better.

Result 2. The converse of Result 1 need not hold. That is, it is possible for the subset model to have lower mse for all predictions. However, this is unusual in practice, as a subset model will have estimated mse lower than the full model only if $F_p < 2/(k' - p)$.

Result 3. If the full model is biased (e.g., some important variables are not included), then there will always be points at which the subset model is better, and points at which the full model is better. In particular, when comparing two subsets, assuming both are biased, neither can have lower mse than the other for all predictions.

These results reemphasize that finding a "best" model for a general purpose such as estimating fitted values will usually be impossible. However, a mse criterion can be used to find several subset models that generally do well relative to the full model, at least over a specified region for predicted values. Most mse criteria can be approximated by use of Mallows' C_p statistic.

Relationship between C_p and F_p. The behavior of the C_p statistic can be illuminated by examining the relationship between C_p and F_p. Since both C_p and F_p are functions of RSS_p, RSS_k, p, k', and n, one can show that

$$F_p = 1 + \frac{C_p - p}{k' - p} \tag{8.16}$$

and, equivalently,

$$C_p = (k' - p)(F_p - 1) + p \tag{8.17}$$

From these, we get the following results for comparing a fixed p parameter model to the full k' parameter model.

1. $C_p < p$ if and only if $F_p < 1$. This is equivalent to the residual mean square criteria (8.13).

2. $C_p < k'$ if and only if $F_p < 2$. $F_p = 2$ is a criterion sometimes used in stepwise regression for adding or deleting variables from a model. This says that a subset model will have C_p smaller than the full model only if $F_p < 2$. Deleting r variables will reduce C_p only if $F_p < 2$.

3. In analogy to the last result, C_p is decreased by adding r variables to a subset model if and only if the F-statistic for testing the significance of those r variables is greater than 2. We would want to add a single variable to a subset model only if $|t| > \sqrt{2}$.

In practice, C_p is used to compare all possible models, or a large class of all possible models in a regression problem. The use of the statistic in this context is discussed in Section 8.6.

8.5   Computational methods I: stepwise methods

The general problem of subset selection requires that the analyst find one model, or a small group of models, from the larger set of possible models such that the selected models are good subsets according to the criteria of interest. A problem arises, however, because the number of possible models is often so large that it is impractical to actually compute a criterion statistic for each model unless one pays careful attention to the method of computation.

Fundamentally, there are two types of techniques that can be used to select subsets. The best methods, discussed in the next section, are used to search through all possible models, or some proportion of them, to find those models that optimize or nearly optimize a criterion. The other important method is called stepwise regression, and has a fundamentally different philosophy. Rather than seek the best subsets, the stepwise procedures provide a systematic technique for examining at most a few subsets of each size. These techniques essentially choose a path through the possible models, looking first at a subset of one size, and then looking only at models obtained from preceding ones by adding or deleting variables.

There are three basic algorithms for stepwise regression, generally called forward selection (FS), backward elimination (BE), and stepwise (SW). In the FS procedure, variables are added at each step. In BE, variables are eliminated. In SW, a step may be either an addition, an elimination, or an interchange of an "in" variable and an "out" variable.

The FS procedure works as follows. First, begin with the simple regression model with that single independent variable that has the biggest sample correlation with the dependent variable Y. This is the first step. Next, we add to the model that independent variable that meets three equivalent criteria: (1) it has the highest sample partial correlation with the dependent variable, adjusting for the independent variables in the equation already; (2) adding the variable will increase R^2 more than any other single variable; and (3) the variable added would have the largest t- or F-statistic of any of the variables that are not in the model. Thus in FS, we start with a subset of size 1, and, at each step, we add another variable to the model, such that the variable added meets the criterion. We continue adding one variable at a time, until a stopping rule is met. The possible stopping rules are:

FS.1 Stop with a subset of a predetermined size p^*.

FS.2 Stop if the F-test for each of the variables not yet entered would be less than some predetermined number, say F–IN. (or, equivalently, stop if the absolute value of the t-statistic would be less than $(F$–IN$)^{1/2}$).

FS.3 Stop when the addition of the next independent variable will make the set of independent variables included in the model too close to collinear. Suppose, for example, that a model currently includes X_1, X_2, and X_3. We will fail to add X_4 to the model if X_4 is nearly a linear combination of X_1, X_2, X_3. The usual procedure is to compute the regression of X_4 on X_1, X_2, X_3 and then compute $1 - R^2$ for this regression. The quantity $1 - R^2$ in this context is called the *tolerance*. If the tolerance is less than some minimum value (usually 0.01 or 0.001, depending on the program), then X_4 is not added to the model. However, one can construct examples for any fixed tolerance, so that addition of a variable that passes a tolerance test will lead to numerically unstable computations (Berk, 1977). Thus the tolerance test should not be viewed as an acceptable substitute for other checks on collinearity. The tolerance check serves two purposes, as it allows the user to identify some collinearities in the data, and it also protects against some round-off error.

In practice, FS methods will use a combination of the above rules for stopping. The usual order is FS.3, then FS.1, and finally FS.2.

The backward elimination (BE) method is similar to FS, except that we start with the full model, and, at each step, remove one variable. The variable to be removed is chosen to be the one that has the smallest t or F value of all the variables in the equation (this is equivalent to removing the variable that causes the smallest change in R^2, or has the smallest partial correlation with Y adjusting for all the other variables left in the model). The stopping rules for BE are also similar to FS:

BE.1   Stop with subset of predetermined size p^*.

BE.2   Stop if the F-test for all the variables now in the model is bigger than some predetermined number, say F–OUT.

The tolerance check is not relevant to BE, since, if the tolerance for some variable were too small, the estimates for the full model could not have been computed. Thus in BE, before we begin, enough variables are deleted so that all of the remaining variables pass the tolerance test. The usual rule for BE is to use either BE.1 or BE.2.

The stepwise algorithm is a combination of FS and BE. At each step, we consider four alternatives: add a variable, delete a variable, exchange two variables, or stop. The rules for SW can be summarized as:

SW.1   If there are at least two variables in the current model, and one or more has a value of F less than F–OUT, the variable with the smallest value of F is removed from the model.

SW.2   If there are two or more variables in the model, the one with the smallest F-value is removed if its removal results in a value of R^2 that is larger than the R^2 obtained for the same number of variables previously. This can happen in the FS procedure since variables are added and deleted at the various steps.

SW.3 If two or more variables are in the model, one of them will be exchanged with a variable not in the model if the exchange increases R^2.

SW.4 A variable is added to the model if it has the highest F-value as in FS, provided that F is greater than F–IN and the tolerance criterion is satisfied.

Using these four rules, several variants of the SW procedure can be defined, depending on how the rules are employed. For example, the program BMDP2R (Dixon and Brown, 1979) allows for four distinct variants: F (rule SW.1 followed by SW.4), $FSWAP$ (SW.1, then SW.3, then SW.4), R (rule SW.2, then SW.4) and $RSWAP$ (SW.2, then SW.3, then SW.4). The methods F and $FSWAP$ may result in more than one subset of size p being considered. With the rules R and $RSWAP$, once a subset of a certain size is attained, we never again look at smaller subsets, but other subsets of the same size may be considered. This algorithm stops when none of the rules can be satisfied, or if the tolerance value is too small.

Choice of parameters F–IN, F–OUT, and tolerance. The tolerance should be set sufficiently low so that computing will stop only if there is in fact a linear dependence in the data or if the resulting estimates will be strongly influenced by round-off error. The exact determination of the tolerance depends on the accuracy of the computer being used and on the number of significant digits in the data, but the value 0.001 is often appropriate.

The choice of F–IN and F–OUT will largely determine the character of the stepwise algorithm. Consider first the FS method. If F–IN is chosen to be very small, say 0.01, the last step of FS will often include all or nearly all of the variables, since only those that fail the tolerance test or are very poor predictors will be excluded. Some analysts use a very small value of F–IN, and then compare the models produced at each step to find a best among this small set of models. This practice is not recommended. The value of F–IN = 2 will be used in the example here because of the analogy between this value and the C_p statistic. Another popular choice of F–IN is 4.0, which will roughly correspond to the 5% level of the F-distribution. On theoretical grounds, Kennedy and Bancroft (1971) suggest using the 25% point of the F-distribution as F–IN, giving a value for F–IN between 2 and 4.

For backward elimination, the choice of a large value of F–OUT will generally have the same effect as the choice of a small value of F–IN. Kennedy and Bancroft suggest using the 10% point of the appropriate F-distribution for F–OUT.

Highway accident data (continued). We shall now apply the various algorithms to the highway accident data given in Table 8.1. We must specify various parameters, such as F–IN, F–OUT, p^*, and the tolerance. For the exposition here, we choose F–IN $= F$–OUT $= 2.0$, and set the tolerance to 0.001 (we do not set p^*). Default values for these parameters vary from program to program. First we use the FS method.

The initial task in FS is to find the single variable which is most highly correlated with the response. From Table 8.2 the correlation between Y and ACPT is biggest, with a value of 0.8075, so this becomes the first variable in the model. The regression is given in the first column of Table 8.5. The partial correlations of RATE adjusted for ACPT with the other potential predictors vary from the smallest (in absolute value) of 0.015 with FA to a maximum of -0.45 with LEN. Thus LEN is added next to the model as shown in Table 8.5. In this manner, we add, in order, SLIM, SIGS, and then PA. The next variable considered is TRKS; however, we see that TRKS has $t = 0.95$, which is less than the $(F$–IN$)^{1/2}$. Hence we stop at step 5 before adding TRKS. If we had used a different value of F–IN, we might have ended with a different model; for example, if $(F$–IN$)^{1/2}$ had been 1.5 instead of 1.414, SIGS would not have been entered, and we would have stopped with the three-variable model.

Table 8.5   FS method

Variable	\multicolumn{6}{c}{Estimate and (t-Values) at Step Number}
	1	2	3	4	5	6
Intercept	1.98	3.19	9.325	8.81	9.94	10.56
	(5.64)	(6.21)	(3.56)	(3.38)	(3.85)	(3.96)
ACPT	0.160	0.145	0.101	0.089	0.064	0.0628
	(6.94)	(6.71)	(3.72)	(3.17)	(2.12)	(2.07)
LEN		0.079	-0.077	-0.0685	-0.074	-0.0635
		(-2.99)	(-3.10)	(-2.72)	(-3.02)	(-2.35)
SLIM			-0.103	-0.096	-0.105	-0.103
			(-2.39)	(-2.26)	(-2.54)	(-2.49)
SIGS				0.485	0.797	0.701
				(1.42)	(2.16)	(1.83)
PA					-0.774	-0.743
					(-1.89)	(-1.80)
TRKS						-0.089
						(-0.95)
d.f.	37	36	35	34	33	32
$\hat{\sigma}^2$	1.76	1.45	1.28	1.24	1.16	1.16
R^2	0.56	0.65	0.70	0.72	0.74	0.75

In the computations shown, the tolerance was checked at each step.

For the BE algorithm, start with the full model, and compute the regression. The variable with the smallest $|t|$ or F is deleted, provided that $|t| < (F\text{–OUT})^{1/2}$ (or $F < F\text{–OUT}$). Then the regression using the remaining variables is computed, and again the variable with smallest $|t|$ or F is deleted. The process is repeated until a stopping criterion is satisfied. The results for BE (or for any stepwise algorithm) are often summarized in a table like Table 8.6 using $F\text{–OUT} = 2.0$. The final model is the same as was found via FS.

Table 8.6 BE for highway data

Step	Removed	R^2	Decrease in R^2	F of Removed Variable
0		0.7605		
1	LANE	0.7604	0.0001	0.01
2	ADT	0.7604	0.0000	0.01
3	SHLD	0.7603	0.0001	0.01
4	FAI	0.7592	0.0012	0.14
5	LWID	0.7579	0.0012	0.15
6	ITG	0.7562	0.0017	0.22
7	MA	0.7521	0.0041	0.52
8	TRKS	0.7450	0.0070	0.90
9	PA	0.7521	0.0071	3.57

Discussion of stepwise procedures. The main advantages of stepwise procedures are that they are fast, easy to compute, relatively inexpensive, and available on virtually all computers and in all computer packages. Also, the cost of the computations increases slowly as the number of variables increases. The FS procedure has the advantage that one can begin with more variables than cases and still arrive at a "solution" to the regression problem. Finally, many of the major computing packages add other options, such as forcing in of variables, addition of variables in groups, or giving some variables priority over others that make stepwise algorithms even more powerful.

Unfortunately, there are important drawbacks to the use of stepwise procedures. Firstly, we reiterate that the model chosen by stepwise regression need not be the best of any criterion of interest; indeed, because of the nature of the one-at-a-time philosophy of stepwise methods, there is no guarantee that the model chosen will in fact include any of the variables that would be in the best subset. Stepwise methods are best when the independent variables are nearly uncorrelated, the condition under which finding a subset model is least likely to be relevant. Also, it is possible to construct examples in which the best subset of size $p = 2$ is completely disjoint from the best subset for $p = 3$, and so on.

Probably the worst indictment of stepwise techniques, at least for the user who is not statistically sophisticated, is that they produce a single result that appears to be *the* model. Similarly, many users pay undue attention to the order in which the variables are entered or deleted from a model. While it is true that the best single variable is entered first in a FS algorithm, there is no guarantee that the best pair is entered as the first pair of variables by the FS algorithm. The ordering of the variables that we get from stepwise regression is an artifact of the algorithm used and need not reflect relationships of substantive interest.

8.6   Computational methods II: all possible regressions

For small problems with only a few independent variables, say $k = 8$ or less, it is not difficult to compute one or two statistics for all possible subsets of variables, and then compare the subsets by comparing their values on the criteria statistics (algorithms are described by Garside (1971), Schatzoff et al. (1968), and by Morgan and Tatar (1972)). In this way, one can easily find a few subsets that are good at explaining the data, and these few can then be subjected to further study (including case analysis, questions of interpretability, usefulness of the results, etc.). If k is large, however, the amount of computing needed to look at all possible subsets appears to grow exponentially as k grows, so if all possible subset regressions are to be computed, a good computational method must be available. An important observation concerning these computations was made by Furnival and Wilson (1974), who suggested that the order of computing regressions could be determined by values of the criteria statistics for models already analyzed. Using this, they wrote an algorithm that, while computing only a small fraction of all possible regressions, will find the best few models on any of a number of criteria. Their algorithm has been implemented in the BMDP series of programs (BMDP9R, Dixon and Brown, 1979) and is also available as subroutine RLEAP in the IMSL library (IMSL, 1979). For k of about 20 or less, the cost of computing involved in the Furnival and Wilson algorithm appears to be about the same as the cost of computing with a stepwise algorithm. Alternative algorithms have been given by Hocking and Leslie (1967), Beale et al. (1967), and LaMotte and Hocking (1970); see also Hocking (1976).

Highway data (continued). As pointed out in Section 8.3, there are 1024 models to consider, 512 with LEN but excluding the 3 dummy variables for highway types, and 512 with LEN and the dummy variables. The procedure we follow is to compute C_p for the best few of these models, using (8.15) with $\hat{\sigma}^2$ always computed from the $k = 13$ variable model given in Table 8.2 (thus in (8.15), $k' = 13 + 1 = 14$). Table 8.7 lists the 20 models with smallest C_p, 10 with the dummy variables and 10 without them. These were found with the Furnival and Wilson algorithm (requiring 0.8 seconds on a CDC Cyber 72 computer). Listed in the table are $p =$ number of parameters, C_p, R^2, the residual sum of squares, and the subscripts of the variables in the model—for example, "1 4 9" includes variables X_1 = LEN, $X_4 =$ SLIM, and $X_9 =$ ACPT.

Table 8.7 20 minimum C_p models

P	C_p	R^2	RSS	Subscripts of Variables in Model
4	0.23	0.70	44.85	1 4 9
5	0.48	0.72	42.33	1 4 8 9
5	0.56	0.72	42.44	1 3 4 9
5	1.33	0.71	43.54	1 4 9 10
5	1.59	0.71	43.91	1 2 4 9
5	1.63	0.71	43.98	1 4 7 9
5	1.97	0.70	44.46	1 4 5 9
6	1.51	0.73	40.93	1 3 4 8 9
6	2.00	0.72	41.64	1 3 4 9 10
6	2.00	0.72	41.64	1 3 4 7 9
7	4.69	0.72	42.62	11 12 13 1 4 9
7	5.12	0.71	43.25	11 12 13 1 4 8
8	3.32	0.75	37.80	11 12 13 1 4 8 9
8	4.68	0.74	39.74	11 12 13 1 3 4 9
9	4.49	0.76	36.54	11 12 13 1 3 4 8 9
9	5.10	0.75	37.47	11 12 13 1 4 7 8 9
9	5.12	0.75	37.50	11 12 13 1 4 6 8 9
9	5.25	0.75	37.68	11 12 13 1 2 4 8 9
9	5.25	0.75	37.69	11 12 13 1 4 5 8 9
9	5.30	0.75	37.76	11 12 13 1 4 8 9 10

All of the minimum-C_p models include LEN (which was forced into all models) and SLIM, while ACPT is included in most models. Adding any one of TRKS, SIGS, ITG, or LANE might have some useful effect, but the need for more than one of them is unlikely.

Examination of Table 8.7 makes clear the fact that there is no model that is obviously best, and there are many equally good models. To make a decision as to which model should be adopted, further analysis is clearly indicated. However, SLIM and ACPT in addition to LEN are certainly useful. Also, the importance of the dummy variables can be judged by an F-test for these variables adjusted for SLIM, ACPT, and LEN—since $F = [(44.85 - 42.62)(32)]/[(42.62)(3)] = 0.56$ with $(3, 32)$ degrees of freedom, there is little evidence that the dummy variables are important predictors of accident rate.

A reasonable approach to finding a model is to begin with the model using only LEN, SLIM, and ACPT as predictors. A case analysis (left to the interested reader) is then in order. The regression summary for this model is given in Table 8.8. One interesting feature of this model is the sign of the estimated coefficient for SLIM—higher speed limits (before the 55 mph maximum limit) were associated with lower accident rates. It would be easy—and incorrect—to assert that higher speed driving lowers accident rates. Rather, one response of a highway department to higher accident rates is to lower speed limits. Thus high accident rates may cause lower speed limits, not the other way around.

Table 8.8   Regression summary

Variable	Estimate	Standard Error	t-Value
Intercept	9.32	2.617	3.56
LEN	− 0.0771	0.0249	− 3.10
SLIM	− 0.102	0.0429	− 2.39
ACPT	0.101	0.0276	3.72

$\hat{\sigma}^2 = 1.281$, d.f. $= 35$, $R^2 = 0.70$

To continue the model-building process, we need to consider adding the other potential variables to the model given in Table 8.8. A graphical method for this study is outlined at the beginning of Chapter 2. For example, the effect of adding SIGS to the model including LEN, SLIM, and ACPT is shown in Figure 8.2. In this plot, the solid line has slope $\hat{\beta}_8 = 0.485$, the slope that would be estimated if SIGS were added to the model. From the graph, it appears that little systematic information will be gained if SIGS is added to the model (the t-value for SIGS in the four-variable model of $t = 1.42$ supports this conclusion). Thus we choose not to include SIGS in the model.

This plot can be a useful diagnostic (in concert with the examination of estimates, t-tests, etc.) for adding variables. However, it is relatively difficult to use since two sets of residuals must be computed, saved, and plotted. Approximations to this plot have been suggested, and are called partial residual plots (Larsen and McCleary, 1972) or residual plus component plots (Wood, 1973).

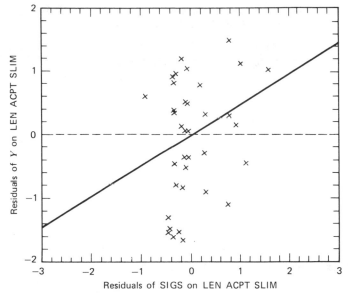

Figure 8.2 Effects of adding SIGS to the model including LEN, SLIM, and ACPT.

The final model. There is no final model, only a group of possible models that are all nearly equally useful. If a model is to be used for prediction, we may be indifferent to the exact variables included. Under these circumstances, the parsimony principle—less is better—can often be applied. In other problems, the analyst's knowledge of a problem must be depended upon to reach conclusions.

Problems

8.1 Using the "data" below, apply the BE and FS algorithms. Also, find C_p for all possible models, and compare results. What is the "correct model"? (Mantel, 1970).

Y	X_1	X_2	X_3
5	1	1004	6.0
6	200	806	7.3
8	− 50	1058	11.0
9	909	100	13.0
11	506	505	13.1

8.2 In the highway accident data, for the fit of the full model, estimate the effect on accident rate of changing a highway classification (1) from MC to MA, (2) MA to PA, and (3) PA to FAI.

8.3 Perform a case analysis on the highway accident data. Are transformations needed? If so, transform the data and find the models with lowest C_p.

8.4 Prove (8.16) and (8.17).

8.5 Use (8.17) to obtain a test of NH: $J_p \leqslant p$ versus AH: $J_p > p$.

8.6 Find equations that (1) give F_p as a function of R_p^2 and (2) give C_p as a function of R_p^2. If a model has the largest R_p^2 for fixed p, does it necessarily have the smallest C_p for that p? Largest $\bar{R}_p^2$ for that p?

8.7 Find C_p for a full model ($p = k'$).

8.8 For the cloud-seeding data of Example 7.1, find the five models with minimum C_p.

8.9 For the boys in the Berkeley Guidance Study (Problem 2.1) find a model for SOMA as a function of the other variables. Perform a complete analysis, including case analysis, and summarize your results.

8.10 Using the methodology outlined by Schatzoff et al. (1968), write a program to compute C_p for all possible regressions.

8.11 An experiment was conducted to model oxygen uptake (O2UP), in

Table 8.10   Data from oxygen uptake experiment (Moore, 1975)

Day	X_1: BOD	X_2: TKN	X_3: TS	X_4: TVS	X_5: COD	O2UP	$Y = \text{Log(O2UP)}$
0	1125.	232.	7160.	85.9	8905.	36.0	1.5563
7	920.	268.	8804.	86.5	7388.	7.9	0.8976
15	835.	271.	8108.	85.2	5348.	5.6	0.7482
22	1000.	237.	6370.	83.8	8056.	5.2	0.7160
29	1150.	192.	6441.	82.1	6960.	2.0	0.3010
37	990.	202.	5154.	79.2	5690.	2.3	0.3617
44	840.	184.	5896.	81.2	6932.	1.3	0.1139
58	650.	200.	5336.	80.6	5400.	1.3	0.1139
65	640.	180.	5041.	78.4	3177.	0.6	-0.2218
72	583.	165.	5012.	79.3	4461.	0.7	-0.1549
80	570.	151.	4825.	78.7	3901.	1.0	0
86	570.	171.	4391.	78.0	5002.	1.0	0
93	510.	243.	4320.	72.3	4665.	0.8	-0.0969
100	555.	147.	3709.	74.9	4642.	0.6	-0.2218
107	460.	286.	3969.	74.4	4840.	0.4	-0.3979
122	275.	198.	3558.	72.5	4479.	0.7	-0.1549
129	510.	196.	4361.	57.7	4200.	0.6	-0.2218
151	165.	210.	3301.	71.8	3410.	0.4	-0.3979
171	244.	327.	2964.	72.5	3360.	0.3	-0.5229
220	79.	334.	2777.	71.9	2599.	0.9	-0.0458

Table 8.11   All possible regressions with log(O2UP) as dependent variable

p	C_p	R^2	RSS	Subscript of Variables in Model
2	6.294	0.6965	1.537	3
2	6.574	0.6926	1.556	5
2	13.500	0.5983	2.034	1
2	20.327	0.5054	2.504	4
2	56.843	0.0082	5.022	2
3	1.739	0.7857	1.085	3 5
3	5.273	0.7376	1.329	4 5
3	6.870	0.7158	1.439	2 5
3	6.883	0.7157	1.440	1 3
3	7.163	0.7119	1.459	3 4
3	7.333	0.7095	1.471	2 3
3	7.704	0.7045	1.496	1 5
3	9.094	0.6856	1.592	1 2
3	11.329	0.6551	1.746	1 4
3	21.365	0.5185	2.438	2 4
4	2.319	0.8050	0.987	2 3 5
4	3.424	0.7900	1.063	3 4 5
4	3.439	0.7898	1.064	1 3 5
4	5.665	0.7595	1.218	2 4 5
4	6.251	0.7515	1.258	1 2 3
4	6.514	0.7479	1.276	1 2 5
4	7.152	0.7392	1.320	1 4 5
4	8.154	0.7256	1.389	1 3 4
4	8.163	0.7255	1.390	2 3 4
4	8.680	0.7184	1.426	1 2 4
5	4.001	0.8094	0.965	2 3 4 5
5	4.318	0.8050	0.987	1 2 3 5
5	5.068	0.7948	1.039	1 3 4 5
5	6.776	0.7716	1.157	1 2 4 5
5	7.695	0.7591	1.220	1 2 3 4
6	6.000	0.8094	0.965	1 2 3 4 5

milligrams of oxygen per minute, from five chemical measurements: biological oxygen demand (BOD), total Kjeldahl nitrogen (TKN), total solids (TS), total volatile solids (TVS), which is a component of TS, and chemical oxygen demand (COD), each measured in milligrams per liter (Moore, 1975). The data were collected on samples of dairy wastes kept in suspension in water in a laboratory for 220 days. All observations were on the same sample over time. We desire an equation relating log(O2UP) to the other variables. The goal is to find variables that should be further studied with the eventual goal of developing a prediction equation (Day cannot be used as a predictor). The data are given in Table 8.10. Table 8.11 gives, for all possible regressions, the summary statistics C_p, R^2, RSS.

A convenient summary for the C_p statistics when the total number of variables considered is small (here, $k = 5$) is called a C_p plot, in which the values of $C_p - p$ are plotted against p (C_p plots are often recommended as just C_p versus p, but the method suggested here makes interpretation somewhat easier). Good models will have $C_p - p$ small—generally less than zero.

8.11.1. Draw a C_p plot. Find the model with minimum C_p. Identify all models for which the F-value for the variables not included is less than 2. Summarize the results.

8.11.2. Complete the analysis of these data, including a complete case analysis. What diagnostic would have indicated the need for transforming O2UP to a logarithmic scale?

9

PREDICTION

Predictions are obtained by substituting values for the independent variables into an estimated regression equation. We then proceed almost as if the estimated prediction function represented a true relationship between the variables. For example, suppose a heavy object is dropped from a tall building. The predicted distance the object will fall in t seconds is $4.9t^2$ meters. As long as t is small enough or the building is high enough, the prediction will be quite accurate as the relationship between distance and time is well known. The parameter 4.9 (meters per second per second), half of the acceleration due to gravity, is a known constant. Predictions of distance made with this equation may not agree perfectly with actual observed values because of measurement errors or neglected factors such as friction.

If the acceleration due to gravity were unknown, predictions could be obtained by first collecting data to estimate it, perhaps by dropping objects and measuring the distance they fall in various times. Letting d denote distance and t denote time, we could fit the model $d = \gamma t^2$ to observed data, and estimate the unknown value of γ. Predictions made from equations with estimated parameters will be more variable due to uncertainty in the estimates, but as long as the functional form of the equation is correct, they will be nonetheless reliable.

In many regression problems, the correct functional form for a prediction equation is unknown, and some predictions (not all) will be useless regardless of how well any parameters are estimated. We rely on the fact that over limited ranges of values for the predictors, many functional forms will behave in nearly the same way; in particular, linear functions

203

(possibly of transformed variables) may provide adequate approximations (see Figure 9.1). In the falling object example, if the true functional form were not known to us, we might decide, using the methods in this book, to fit a linear model using perhaps log(time) as a predictor. The resulting predictions can be expected to be reasonably good as long as we predict for values of t roughly in the same range as the values of t in the sample used for model building, because, over modest ranges for t, the correct functional form $d = \gamma t^2$ can be approximated by $d = \beta_0 + \beta_1 \log(t)$.

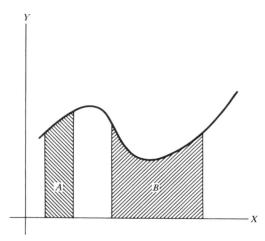

Figure 9.1 A linear approximation in range A may be adequate and a quadratic approximation may be reasonable for B. Neither would be adequate outside of the range of fitting.

If a functional relationship between variables is known, the distinction between estimation of parameters and prediction is blurred, since the parameters may have meaningful interpretations, such as the acceleration due to gravity discussed previously. However, if the form of the prediction function must be estimated, the difference between estimation and prediction is both meaningful and instructive. We estimate parameters with the purpose of obtaining predicted values. The parameters themselves will rarely have intrinsic meaning. If we fit $\hat{d} = \hat{\beta}_0 + \hat{\beta}_1 \log(t)$ to a set of falling object data, β_0 and β_1 are artifacts of the equation fit, not of the relationship between time and distance. The estimated values $\hat{\beta}_0$ and $\hat{\beta}_1$ depend on the range of the dependent variable in the data, so the parameters β_0 and β_1 represent quantities that depend on the data, not on the relationship between the variables (they are "variable constants"). Even so, predictions derived from the fitted model may be useful.

9.1 Making predictions

Where do the data come from? In the ideal case for prediction, data for estimation of the prediction function and future cases for which predictions are desired are a random sample from some well-defined population. Then, usual modes of statistical inference from a sample to a population apply. In most practical studies, these circumstances are an idealization that is never exactly true. For example, in a chemical experiment cases may be obtained by repeating a procedure with varying combinations of independent variables. If the experiment were done in one day, or by one technician, or in one lab, or with one batch of a chemical, there is no general principle to guarantee that the estimated prediction equation will apply to experimental results from a different day, technician, or lab. One must either assume that the effect due to these factors is negligible, or else collect data from several technicians or labs to assess their effects.

In other circumstances, the cases used for estimation may not represent the population for which predictions are desired. Often, the cases used in estimation may be drawn from a subset of the population of interest. In a dietary study, for example, if only healthy rats are used to collect data, generalization to a population of all rats may not be possible, as the data used in estimating the prediction function only represent a subset of the target population.

Where does the functional form come from? The true functional form that relates an independent variable to a set of predictors is rarely known, and often the equation that is used must be empirically determined. We may choose a functional form to satisfy criteria that are numerical, such as C_p, or graphical, such as giving good residual plots. The functional form chosen optimizes the fit to the data, taking all the oddities of the observed numbers into account. The fitted equation tends to match the observed data better than the true relationship would if it were known. Consequently, predictions for other cases will tend to be somewhat worse than would be estimated from the statistics computed in the analysis.

Interpolation versus extrapolation. If the functional form is either estimated from the data analysis, or if it is suspected that the form used is valid only for some values of the predictors, then some predictions will be reliable and others will not. In a chemical experiment, for example, changing to an extreme temperature may make the actual outcomes completely different from those predicted from experiments at moderate temperatures. Consequently, in most prediction problems there is an important question of determining a range of validity for predictions. Cases with dependent variables within the range of validity are termed *interpola-*

tions, while cases outside the range are called *extrapolations*. Interpolations tend to be accurate and reliable, while extrapolations are less accurate and possibly unreliable. The question of determining the range of validity for predictions is discussed in Section 9.3.

Making predictions. We shall need some additional notation before proceeding. Suppose we have n cases, each with p independent variables and a response measured, to be used to estimate a prediction function. The observed data are called the construction sample. Now consider a new case for which a prediction is to be made. Values of variables for the case to be predicted will have a tilde ($\tilde{\ }$) above them, so that $\tilde{\mathbf{x}}^T = (1, \tilde{x}_1, \tilde{x}_2, \ldots, \tilde{x}_p)$ is the vector of predictor variables; if more than one prediction is to be made, add a subscript k for the kth prediction. Also, let $\tilde{y}$ denote the predicted response for $\tilde{\mathbf{x}}$. Unweighted least squares will be used for estimation (if weighted least squares is being used, first apply a transformation to reduce the problem to unweighted least squares). Fit the linear model $\mathbf{Y} = \mathbf{X}\boldsymbol{\beta} + \mathbf{e}$, $\text{cov}(\mathbf{e}) = \sigma^2\mathbf{I}$ so that the estimate of $\boldsymbol{\beta}$ is $\hat{\boldsymbol{\beta}} = (\mathbf{X}^T\mathbf{X})^{-1}\mathbf{X}^T\mathbf{Y}$, where $\mathbf{X}$ ($n \times p'$) and $\mathbf{Y}$ ($n \times 1$) are both from the construction sample and do not involve the case to be predicted. The prediction function, relating the point prediction $\tilde{y}$ to $\tilde{\mathbf{x}}$ is then given by

$$\tilde{y} = \tilde{\mathbf{x}}^T\hat{\boldsymbol{\beta}} = \hat{\beta}_0 + \hat{\beta}_1\tilde{x}_1 + \hat{\beta}_2\tilde{x}_2 + \cdots + \hat{\beta}_p\tilde{x}_p \qquad (9.1)$$

If the linear model is correct, then the prediction function will yield an unbiased point estimate of the true value, given by $\tilde{\mathbf{x}}^T\boldsymbol{\beta} + \tilde{e}$, where we include the term $\tilde{e}$ to reflect the unmodeled variability in the response. As in the construction sample, we assume $\tilde{e}$ has zero mean and variance σ^2. The variance due to estimating $\tilde{\mathbf{x}}^T\boldsymbol{\beta}$ by $\tilde{\mathbf{x}}^T\hat{\boldsymbol{\beta}}$, is $\text{var}(\tilde{\mathbf{x}}^T\hat{\boldsymbol{\beta}}) = \sigma^2[\tilde{\mathbf{x}}^T(\mathbf{X}^T\mathbf{X})^{-1}\tilde{\mathbf{x}}]$ $= \sigma^2\tilde{v}$, where $\tilde{v} = \tilde{\mathbf{x}}^T(\mathbf{X}^T\mathbf{X})^{-1}\tilde{\mathbf{x}}$; compare to the definition of v_{ii} given by (5.8). The important difference between $\tilde{v}$ and v_{ii} is that, in the definition of v_{ii}, the vector $\mathbf{x}_i^T$ was one of the rows of $\mathbf{X}$; here, $\tilde{\mathbf{x}}^T$ need not be a row of $\mathbf{X}$. Consequently, while the v_{ii}'s are bounded between 0 and 1, $\tilde{v}$ may take on any positive value (or, if the intercept is in the model, any positive value greater than $1/n$). The greater $\tilde{v}$ is, the larger the variance is, and the farther $\tilde{\mathbf{x}}$ is from the data in the construction sample.

Combining the two sources of variability, we get the variance of prediction, varpred,

$$\text{varpred}(\tilde{y} \mid \tilde{\mathbf{x}}) = \sigma^2(1 + \tilde{v}) \qquad (9.2)$$

This is a very reasonable form for the prediction variance. First, no matter how well we know $\tilde{\mathbf{x}}^T\boldsymbol{\beta}$ (i.e., $\tilde{v}$ is small) the variance of prediction is never less than σ^2, the residual variance. Secondly, if our knowledge of $\tilde{\mathbf{x}}^T\boldsymbol{\beta}$ is poor ($\tilde{v}$ large), our predictions can be much less precise than residual variability. These will correspond to interpolation and extrapolation, respectively.

In practice, σ^2 is not known, so the residual mean square $\hat{\sigma}^2$ from the fitted model is substituted into (9.2) to get the estimated prediction variance. The square root of this quantity is called the standard error of prediction, written sepred($\tilde{y} \mid \tilde{x}$), and is given by

$$\text{sepred}(\tilde{y} \mid \tilde{x}) = \hat{\sigma}\sqrt{1 + \tilde{v}} \tag{9.3}$$

Example 9.1 **Predicting the interval to eruption of Old Faithful geyser.**

A *geyser* is a hot spring that occasionally becomes unstable and erupts hot water and steam into the air. Different geysers will erupt for various lengths of time and intervals. Probably the most famous geyser is called Old Faithful, in Yellowstone National Park in Wyom-

Table 9.1 **Old Faithful geyser data, X = duration of eruption (min), Y = interval between eruptions (min)**

Date	X	Y	Date	X	Y
August 1	4.367	74.000	August 3	4.500	71.000
	3.867	70.000		3.917	78.000
	4.000	64.000		4.350	80.000
	4.033	72.000		2.333	51.000
	3.500	76.000		3.833	82.000
	4.083	80.000		1.883	49.000
	2.250	48.000		4.600	80.000
	4.700	88.000		1.800	43.000
	1.733	53.000		4.733	83.000
	4.933	71.000		1.767	49.000
	1.733	56.000		4.567	75.000
	4.617	69.000		1.850	47.000
	3.433	72.000		3.517	78.000
August 2	4.250	76.000	August 4	4.000	71.000
	1.667	54.000		3.700	69.000
	3.917	76.000		3.717	63.000
	3.683	65.000		4.250	64.000
	3.100	54.000		3.583	82.000
	4.033	86.000		3.800	68.000
	1.767	40.000		3.767	71.000
	4.083	87.000		3.750	71.000
	1.750	49.000		2.500	63.000
	3.200	76.000		4.500	79.000
	1.850	51.000		4.100	66.000
	4.617	77.000		3.700	75.000
	1.967	49.000		3.800	56.000
				3.433	83.000

Table 9.1   (continued)

Date	X	Y	Date	X	Y
August 5	4.000	67.000	August 7	3.500	50.000
	2.267	65.000		1.967	87.000
	4.400	77.000		4.283	40.000
	4.050	72.000		1.833	76.000
	4.250	79.000		4.133	57.000
	3.333	73.000		1.833	71.000
	2.000	53.000		4.650	70.000
	4.333	69.000		4.200	69.000
	2.933	53.000		3.933	72.000
	4.583	78.000		4.333	51.000
	1.900	55.000		1.833	84.000
	3.583	67.000		4.533	43.000
	3.733	68.000	August 8	4.183	73.000
	3.733	73.000		4.433	73.000
August 6	1.817	53.000		4.067	70.000
	4.633	70.000		4.133	84.000
	3.500	69.000		3.950	71.000
	4.000	66.000		4.100	79.000
	3.667	79.000		2.717	58.000
	1.667	48.000		4.583	73.000
	4.600	90.000		1.900	59.000
	1.667	49.000		4.500	76.000
	4.000	78.000		1.950	49.000
	1.800	52.000		4.833	75.000
	4.417	79.000		4.117	75.000
	1.900	49.000			
	4.633	75.000			
	2.933	75.000			

ing. The interval between eruptions of Old Faithful ranges between about 30 and 90 minutes, to heights generally over 35 meters with eruptions lasting from 1 to $5\frac{1}{2}$ minutes (Marler 1969). Because of its regularity and beauty, Old Faithful geyser has become a major tourist attraction, and it is not uncommon on summer afternoons for several thousand people to watch an eruption.

Prediction of the time of the next eruption of this geyser is of interest both to the National Park Service and to visitors. In fact, prediction of the time of the next eruption is posted in the Old Faithful Geyser Visitor's Center. The data given in Table 9.1 lists X = the duration of an eruption and Y = interval to the next eruption (both in minutes) for eruptions of Old Faithful geyser between 6 AM and Midnight for August 1 to August 8, 1978. The data were collected by the ranger/naturalists at Old Faithful Geyser Visitor Center, and recorded by hand in a log book. Measurements were made using stop watches.

The National Park Service uses values of X to predict future values for Y, and we shall follow its example. It is unlikely that a causal relationship exists between X and Y, rather, both X and Y are probably results of some other unobserved cause or set of causes. Modeling the observed association between X and Y may lead to useful predictions, but the fitted relationship may have no geological meaning.

The scatter plot of Y versus X for the 106 cases in the data is given in Figure 9.2. The plot generally appears to show that shorter eruptions are followed by shorter intervals, and longer eruptions are followed by longer intervals. The points appear to fall into two clusters in the lower-left- and upper-right-hand corners, with a few stragglers elsewhere (more will be made of this later). As a first approximation, a simple regression model

$$Y = \beta_0 + \beta_1 X + \text{error} \tag{9.4}$$

can be fit to the data. Summary statistics for the fit of (9.4) are given in Table 9.2. Since $R^2 = 0.36$, knowledge of X does help predict Y. The fitted equation is

$$\tilde{y} = 42.5 + 7.2\tilde{x} \tag{9.5}$$

where $\tilde{x} =$ duration of the last eruption, and $\tilde{y}$ is the predicted interval to the next eruption. Sepred($\tilde{y} \mid \tilde{x}$) can be computed from

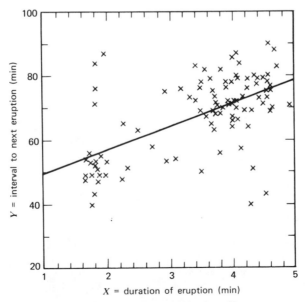

Figure 9.2   Old Faithful geyser data.

Table 9.2   Old Faithful geyser data

Variable	n	Summary statistics Sample Mean	Sample Variance	Minimum	Maximum
X	106	3.474	1.074	1.667	4.933
Y	106	67.42	152.6	40.00	90.00

Variable	Regression of Y on X Estimate	Standard Error	t-Value
Intercept	42.51188	3.384589	12.56
X	7.169077	0.934042	7.68

$\hat{\sigma}^2 = 9.91885$, d.f. $= 104$, $R^2 = 0.3616$.

equation (9.2), where $\tilde{v} = 1/n + (\tilde{x} - \bar{x})^2/SXX$, with $\bar{x}$ = average of X's in the data and SXX defined in Chapter 1. The value of SXX can be computed from the statistics in Table 9.2, since the sample variance of X is $SXX/(n - 1) = 1.074$, so $SXX = (1.074)(106) = 112.77$. Thus

$$\text{sepred}(\tilde{y} \mid \tilde{x}) = 9.92\left(1 + \frac{1}{106} + \frac{(\tilde{x} - 3.47)^2}{112.77}\right)^{1/2} \tag{9.6}$$

For example, if $\tilde{x} = 4.5$ minutes, $\tilde{y} = 42.5 + 7.2(4.5) = 74.8$ minutes,

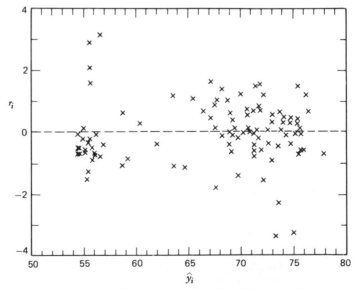

Figure 9.3   Residual plot for Old Faithful geyser data.

and sepred($\tilde{y} \mid \tilde{x} = 4.5$) = 9.92 $(1 + 1/106 + (4.5 - 3.74)^2/112.77)^{1/2}$ 10.01 minutes.

Case analysis is needed to determine whether the simple regression model fit is in fact reasonable and appropriate, and to see if a few of the cases in the data are overly influential in the fitting of the model. One aspect of this analysis is summarized in a plot of r_i versus fitted values (Figure 9.3), and in a rankit plot of the residuals in Figure 9.4. In the residual plot, the clustering of points into two groups persists: if a method of determining which cluster a case belongs to could be determined, improved predictions might result. The rankit plot indicates that the residuals are clearly non-normal, a fact that will make interval predictions difficult.

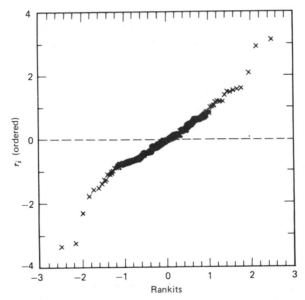

Figure 9.4 Rankit plot for Old Faithful geyser data.

If we view both X and Y as being caused by some other set of variables, then, as predictors of future values of Y we may reasonably use both the current value of X and any of the values of X and Y observed in the past. For example, we can model Y, the interval to the next eruption, as

$$Y = \beta_0 + \beta_1 X + \beta_2 \text{ (duration of last eruption)}$$
$$+ \beta_3 \text{ (interval between last two eruptions)} + \text{error} \qquad (9.7)$$

These two additional variables are called *lagged variables*.

To fit model (9.7), the data in Table 9.1 are reduced to 98 cases, since for the first recorded eruption on each day, the lagged interval and duration are unknown. The data for August 1, for example, are

Y	X	X(lagged)	Y(lagged)
70	3.867	4.367	74
64	4.000	3.867	70
72	4.033	4.000	64
⋮	⋮	⋮	⋮
69	4.617	1.733	56
72	3.433	4.617	69

The regression of Y on X, X(lagged), and Y(lagged) is summarized in Table 9.3. It is evident that all three are useful predictors after the others. This model appears to be a substantial improvement over the model with just X as a predictor as indicated by the increase in R^2 from 0.36 to 0.71, and in the residual plot, Figure 9.5. Notice that the clustering apparent in Figures 9.2 and 9.3 is not as pronounced, so addition of the two lagged variables does help classify cases into appropriate clusters. The fitted model is

$$\hat{y} = 79.50 + 6.43X + 5.52X(\text{lagged}) - 0.77Y(\text{lagged}) \quad (9.8)$$

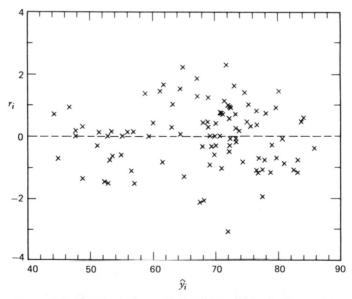

Figure 9.5   Residual plot, with lagged variables in the model.

Table 9.3 Regression with lagged variables

Variable	Estimate	Standard Error	t-Value
Intercept	79.500	6.222	12.78
X	6.434	0.831	7.74
X(lagged)	5.522	0.966	5.72
Y(lagged)	0.793	0.075	-10.59

$\hat{\sigma}^2 = 47.87$, d.f. $= 94$, $R^2 = 0.7067$.

It is interesting that long intervals tend to be followed by shorter intervals, as indicated by the negative coefficient for Y(lagged).

To find standard errors of prediction for this model, equation (9.3) is applied.

9.2 Interval estimates

If the errors both in the construction sample and in cases for prediction are normally distributed, with zero mean and common variance σ^2, then the prediction $\tilde{y}$ (given $\tilde{x}$) is normally distributed, with the mean $\tilde{x}^T\beta$ and variance varpred($\tilde{y} \mid \tilde{x}$). Hence, confidence intervals for $x^T\beta$ can be based on Student's t-distribution, with degrees of freedom equal to the degrees of freedom in $\hat{\sigma}^2$, usually $n - p'$. The $(1 - \alpha) \times 100\%$ confidence interval for $x^T\beta$ is given by the set of all points y such that

$$\tilde{y} - t(\alpha; n - p')\text{sepred}(\tilde{y} \mid \tilde{x}) \leqslant y \leqslant \tilde{y} + t(\alpha; n - p')\text{sepred}(\tilde{y} \mid \tilde{x}) \quad (9.9)$$

If M predictions are to be made for predictor vectors $\tilde{x}_1, \tilde{x}_2, \ldots, \tilde{x}_M$, then simultaneous prediction intervals can be formed (using the Bonferroni method) by replacing $t(\alpha; n - p')$ in (9.9) by $t(\alpha/M; n - p')$.

Hypothesis tests using Student's t-distribution about predictions can be constructed in the usual fashion. The hypothesis that $\tilde{x}^T\beta$ has a particular value, say y^*, against a general alternative is tested by comparing $t = (\tilde{x}^T\hat{\beta} - y^*)/\text{sepred}(\tilde{y} \mid \tilde{x})$ to Student's t with $n - p'$ degrees of freedom.

Old Faithful geyser (continued). For a new case with $\tilde{X} = 4.1$ minutes, $\tilde{X}$(lagged) $= 2.0$ minutes, $\tilde{Y}$(lagged) $= 61$ minutes, the predicted time to the next eruption is

$$\tilde{y} = 79.5 + 6.43(4.1) + 5.52(2.0) - 0.79(61) = 68.5 \text{ minutes}$$

One can show that $\tilde{v} = 0.0331$, and, since $\hat{\sigma} = (47.87)^{1/2} \cong 6.9$, a 95%

prediction interval for the future value is the set of points y such that

$$68.5 - (1.98)(6.9)\sqrt{1 + 0.0331} < y < 68.5 + (1.98)(6.9)\sqrt{1 + 0.0331}$$

or

$$55 \text{ minutes} \leqslant y \leqslant 82 \text{ minutes}$$

The prediction method actually used by the National Park Service corresponds to fitting the regression of Y on X and then rounding the estimated coefficients to convenient multiples. It claims that an interval of ± 5 minutes corresponds to about a 75% prediction interval. However, from (9.6), we see that ± 5 minutes for their method corresponds to about ± 0.5 sepred, or about a 40% prediction interval (assuming normality). Using equation (9.8) leads to ± 5 minutes corresponding to about ± 0.7 sepred, or about a 50% prediction interval. Neither method is equal to the claim made by the Park Service.

Predictions in a different scale. If a fitted linear model gives predictions in a transformed scale, it may occasionally be desirable to change a predicted value back to the original units. For example, in Example 6.1, the relationship between brain and body weight of mammal species was discussed, and the equation

$$\widehat{\log(\text{brain weight})} = 0.9271 + 0.7517 \log(\text{body weight})$$

was found to represent the data well. If a new species had an average body weight of 10 kilograms, the predicted log(brain weight) = 0.9271 + 0.7517 log(10) = 1.6788. A naive estimate of the brain weight of this species might be the antilogarithm $10^{1.6788} = 47.73$ grams.

Unfortunately, this estimate is biased and will be, on the average, too small. The naive practice of reexpressing a transformed predictor will give an estimate of the median of distribution of the predicted value given the predictors, rather than the mean, as would be the case if no transformation were involved. Methods for finding unbiased (or nearly unbiased) estimates of the mean for reexpressed predicted values are complicated, and will depend both on the transformation and on assumptions about the errors in the transformed scale (e.g., normality). For some transformations, such as the logarithmic, a table look-up is required to compute the unbiased estimates.

However, unlike point estimates, confidence statements or predictive intervals are easily converted from one scale to another because estimates of quantiles (percentage points) in one scale are just the transforms of the quantile estimates in another scale. An interval for a prediction is obtained by reexpressing the limits of the computed interval. In the brain weight example, a 95% prediction interval for log(brain weight) of a 10-kilogram mammal species is given by the set $1.0704 \leqslant \log(\text{brain weight}) \leqslant 2.2872$. The corresponding 95% confidence interval for brain weight is $10^{1.0704} \leqslant \text{brain weight} \leqslant 10^{2.2872}$ or between 11.8 grams and 193.7 grams, which is a very wide interval. Note that the interval is not symmetric about the naive predictor $10^{1.6788} = 47.73$ grams.

This method of getting confidence intervals is always correct in that, given the assumptions, the actual confidence level is equal to the stated confidence level, and should therefore be adequate for most purposes. However, the interval will not be optimal in the sense of having the shortest length. As with point predictors, shortest length intervals will usually require a table look-up. For further details, see Land (1974).

Before any reexpression of scale is done, the careful analyst should consider if the new scale is any better than the old. In the brain weight–body weight example, expression of the weights using logarithms may be more natural if in this scaling weight is linearly related to other interesting quantities.

9.3   Interpolation versus extrapolation

Interpolation means prediction for new cases with predictor variables not too different from the values of the predictor variables in the construction sample. For a simple linear regression model, interpolation generally occurs when the predictor is in the range observed in the construction sample; outside of this range, we would call a prediction an extrapolation. In the Old Faithful geyser data, we should be reluctant to make predictions for Y if X is very short (say a few seconds) or very long (say 10 minutes) since, from our data, we have no certain knowledge of the relationship between Y and X under those conditions.

The usefulness of the distinction between interpolation and extrapolation is illustrated in Figure 9.6. If the form of the prediction function is not known a priori, then we have no information on the relationship outside the observed range for the predictor. Any of the dotted paths shown in the figure may be appropriate and, without further information, we cannot know which should be used. For an interpolation to be useful, only a few assumptions must be satisfied, principally that the new case behaves like those in the construction sample. For extrapolation, the assumption that the estimated prediction function is relevant to cases with predictors outside of the interpolatory range is also needed. For example, rainfall generally increases crop yields, and data collected over a reasonable range of rainfalls might give a good prediction function for yield from rainfall. This function, however, might be useless for extrapolations. According to

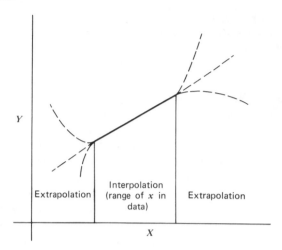

Figure 9.6   Extrapolation.

the equation, zero rainfall might imply negative crop yields and very large rainfalls might result in predictions of very large yields—even though, after a point more rainfall will decrease yield.

In multiple regression, it is difficult to define a range of validity for prediction. Clearly, it must depend on the observed data in the construction sample. Suppose, as an example, we consider the Old Faithful geyser data, except use only X and X(lagged) as the independent variables (this is done so that a two-dimensional picture is possible, but the methodology is perfectly general and can be applied to higher dimensional data). Figure 9.7 gives a scatter plot of X versus X(lagged). A prediction would be called an interpolation if it had $(X, X$(lagged)) like those used to construct the prediction equation. This suggests a possible definition of the range of validity to be the smallest closed figure that includes all the points. We shall consider two approximations to this region, one easy to find, one not so easy.

Recall that contours of constant $v_{ii} = \mathbf{x}_i^T(\mathbf{X}^T\mathbf{X})^{-1}\mathbf{x}_i$ are ellipsoids. If v_{max} is the largest v_{ii} in the construction sample, the set of all points $\tilde{\mathbf{x}}$ such that $\tilde{\mathbf{x}}^T(\mathbf{X}^T\mathbf{X})^{-1}\tilde{\mathbf{x}} \leqslant v_{max}$ is an ellipsoid which completely includes all the observed data in the construction sample. This ellipsoid is drawn on Figure 9.7 as a solid curve. It is centered at point "1" on the graph. For a prediction at point $\tilde{\mathbf{x}}$, if $\tilde{\mathbf{x}}^T(\mathbf{X}^T\mathbf{X})^{-1}\tilde{\mathbf{x}} = \tilde{v} > v_{max}$, then $\tilde{\mathbf{x}}$ is not in the ellipse, and the prediction can be regarded as an extrapolation.

In Figure 9.7, it is clear that the set defined by the solid curve is much larger than the smallest set containing the observed data. To provide a better approximation, the smallest volume ellipsoid containing these points may be found. This is called the minimum covering ellipsoid or MCE (Titterington, 1975; Cook and Weisberg, 1979). The MCE may have a different center and different axes than does the ellipsoid defined using $(\mathbf{X}^T\mathbf{X})$; for the example, the MCE is given by the dotted line in Figure 9.7.

In general, an ellipsoid is determined by a vector $\mathbf{m}$ and a symmetric positive definite matrix $\mathbf{M}$. Once $\mathbf{m}$ and $\mathbf{M}$ are determined for the MCE, checking to see if a vector of independent variables $\tilde{\mathbf{x}}^T = (\tilde{x}_1, \ldots, \tilde{x}_p)$ is contained in the MCE is straightforward (note here the constant 1 is not in $\tilde{\mathbf{x}}$). For models with an intercept, compute

$$(\tilde{\mathbf{x}} - \mathbf{m})^T \mathbf{M}^{-1} (\tilde{\mathbf{x}} - \mathbf{m}) = \tilde{w}$$

If $\tilde{w}$ is greater than p, the number of independent variables in the model, then the prediction at $\tilde{\mathbf{x}}$ is an extrapolation. (For the solid line in Figure 9.7, take $M = \mathfrak{X}^T \mathfrak{X} / n$, and $\mathbf{m} = \bar{\mathbf{x}}$).

An outline of an algorithm for finding the MCE is given in Appendix 9A.

As can be seen in Figure 9.7, the MCE more nearly matches the data than does the usual ellipsoid based on $\mathfrak{X}^T \mathfrak{X}$. However, even this region for interpolation may be too large, since there is a substantial area inside the MCE with no observed data, so the MCE can be considered to be only an approximation to the region of interest.

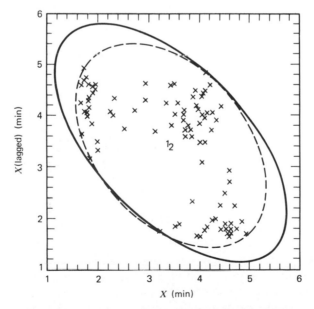

Figure 9.7 Approximations to interpolatory regions.

Problems

9.1   In the Old Faithful geyser data, any possible effects of day-to-day variation have been neglected. How could possible day-to-day variation be included in the model? If there were day-to-day variation, under what circumstances would it make predictions impossible? Under what circumstances would predictions be possible but more variable? Fit a model that takes day-to-day effects into account and summarize results.

9.2   Consider adding lag-two variables (i.e., the values of X and Y are the eruption before last) to model (9.7). Are these variables useful?

9.3   (John Rice). In heart cathetarization a teflon tube (or catheter) of 3-millimeter diameter is passed into a major vein or artery at the femoral region and moved into the heart. The catheter can be maneuvered into specific regions to provide information concerning the physiology and function of the heart. This procedure is sometimes done on children with congenital heart defects where the physician must guess the proper length of the catheter.

In a small study of 12 patients the proper length Y of the catheter was determined by checking with a fluoroscope (X-ray) that the catheter tip had reached the aortic valve. Each patient's X_1 = height (inches) and X_2 = weight (in pounds) were also recorded to see if these could be helpful in predicting catheter length (in centimeters). The data are given below.

Y	X_1	X_2
37	42.8	40.0
50	63.5	93.5
34	37.5	35.5
36	39.5	30.0
43	45.5	52.0
28	38.5	17.0
37	43.0	38.5
20	22.5	8.5
34	37.0	33.0
30	23.5	9.5
38	33.0	21.0
47	58.0	79.0

Construct prediction equations using X_1 alone, X_2 alone, and both X_1 and X_2. If an error in prediction of ± 2 centimeters is tolerable, are any of the prediction equations adequate (you must decide what the "± 2-centimeter" means as a probability statement)? Find the region of applicability for the models fit, and summarize your results.

9.4 Beginning in the middle 1970s, law and other professional schools have had many more applicants for admission than places for students. As a result, equitable methods of choosing between candidates for admission are very important. Regression ideas are applied in the following way. First data are collected on students currently enrolled in the school. Typical measurements taken are undergraduate grade point average (X_1), score on a standardized test such as the Law School Aptitude Test (X_2), and a measure of performance, such as first-year grade point average (Y). Then, a prediction function of the form $\hat{Y} = \hat{\beta}_0 + \hat{\beta}_1 X_1 + \hat{\beta}_2 X_2$ is estimated (although the estimates may not be obtained by least squares; see Rubin, 1980). Now, in the pool of applicants both X_1 and X_2 are available, so a value of $\hat{Y}$ can be computed for each applicant, and applicants with large values of $\hat{Y}$ are admitted (although not all schools decide admission solely by $\hat{Y}$).

Discuss the problems in using this methodology. In particular, compare the population for construction to the target population. Suppose the correlation between X_1 and X_2 for current students turned out to be negative (as it occasionally does). Explain how this could happen. What, if anything, does such a correlation mean? (See also Aitken, 1934 and Lawley, 1943.)

9.5 Property taxes on a house are supposedly dependent on the current market value of the house. Since houses actually sell only rarely, the sale price of each house must be estimated every year when property taxes are set. Regression methods are sometimes used to make up a prediction function (Renshaw, 1958).

The data in Table 9.4 are for $n = 28$ houses in Erie, Pennsylvania that were actually sold (Narula and Wellington, 1977). The variables are:

X_1 = current taxes (local, school, and county) $\div$ 100 (dollars).

X_2 = number of bathrooms.

X_3 = lot size $\div$ 1000 (square feet).

X_4 = living space $\div$ 1000 (square feet).

X_5 = number of garage spaces.

X_6 = number of rooms.

X_7 = number of bedrooms.

X_8 = age of house (years).

X_9 = number of fireplaces.

Y = actual sale price $\div$ 1000 (dollars).

Use the data to estimate a function to predict Y from the X's and functions of the X's. (In practice, the data set used to estimate sale prices would be much larger than the data set available here, and other variables for neighborhood indicators, quality of schools, etc. would be included.)

Table 9.4 House data[a]

X_1	X_2	X_3	X_4	X_5	X_6	X_7	X_8	X_9	Y
4.9176	1.0	3.4720	.9980	1.0	7	4	42	0	25.9
5.0208	1.0	3.5310	1.5000	2.0	7	4	62	0	29.5
4.5429	1.0	2.2750	1.1750	1.0	6	3	40	0	27.9
4.5573	1.0	4.0500	1.2320	1.0	6	3	54	0	25.9
5.0597	1.0	4.4550	1.1210	1.0	6	3	42	0	29.9
3.8910	1.0	4.4550	.9880	1.0	6	3	56	0	29.9
5.8980	1.0	5.8500	1.2400	1.0	7	3	51	1	30.9
5.6039	1.0	9.5200	1.5010	0	6	3	32	0	28.9
15.4202	2.5	9.8000	3.4200	2.0	10	5	42	1	84.9
14.4598	2.5	12.8000	3.0000	2.0	9	5	14	1	82.9
5.8282	1.0	6.4350	1.2250	2.0	6	3	32	0	35.9
5.3003	1.0	4.9883	1.5520	1.0	6	3	30	0	31.5
6.2712	1.0	5.5200	.9750	1.0	5	2	30	0	31.0
5.9592	1.0	6.6660	1.1210	2.0	6	3	32	0	30.9
5.0500	1.0	5.0000	1.0200	0	5	2	46	1	30.0
5.6039	1.0	9.5200	1.5010	0	6	3	32	0	28.9
8.2464	1.5	5.1500	1.6640	2.0	8	4	50	0	36.9
6.6969	1.5	6.9020	1.4880	1.5	7	3	22	1	41.9
7.7841	1.5	7.1020	1.3760	1.0	6	3	17	0	40.5
9.0384	1.0	7.8000	1.5000	1.5	7	3	23	0	43.9
5.9894	1.0	5.5200	1.2560	2.0	6	3	40	1	37.5
7.5422	1.5	4.0000	1.6900	1.0	6	3	22	0	37.9
8.7951	1.5	9.8900	1.8200	2.0	8	4	50	1	44.5
6.0931	1.5	6.7265	1.6520	1.0	6	3	44	0	37.9
8.3607	1.5	9.1500	1.7770	2.0	8	4	48	1	38.9
8.1400	1.0	8.0000	1.5040	2.0	7	3	3	0	36.9
9.1416	1.5	7.3262	1.8310	1.5	8	4	31	0	45.8
12.0000	1.5	5.0000	1.2000	2.0	6	3	30	1	41.0

[a] Source: Narula and Wellington (1977).

10

INCOMPLETE DATA

In many data sets, some variables will be unrecorded for some cases. Indeed, in large studies complete data are more the exception than the rule. Since the standard methods of analysis can be applied directly only to complete data sets, additional techniques are needed. The most common technique requires that the data set be modified, either by deleting partially observed cases or variables, or by filling in guesses for unobserved values. A usual analysis is then performed, adjusting where necessary to account for the modifications made in the data. Alternatively, methods for analysis of incomplete data without filling in or deleting have been developed. Use of these generally relies on strong assumptions. As we shall see, neither of these general approaches is wholly satisfactory.

The statistical literature on incomplete data problems is very large, and the treatment of the subject given here is not comprehensive. A bibliography on the subject is given by Afifi and Elashoff (1966) and in the references given in the more recent papers cited in this chapter.

10.1 Missing at random

Most of the methodology for analyzing incomplete data uses the assumption that the cause of values being unobserved is unrelated to the relationships under study. For example, data that are unobserved because of a dropped test tube or a lost coding sheet would ordinarily satisfy this assumption. Under circumstances like these, analysis that ignores the causes of failure to observe values is justified. On the other hand, if the reason for not observing values depends on the values that would have been observed, then the analysis of data must include modeling the cause of the failure to observe values.

Rubin (1976) has made a precise distinction between the two types of incomplete data. For this discussion, an incomplete data set will have values that are *missing at random* (MAR) if the failure to observe a value does not depend on the value that would have been observed. Determining whether or not an assumption of MAR is appropriate for a particular data set is an important step in the analysis of incomplete data. The following examples illustrate the application of the definition.

Dart-throwing. Imagine that each of the numbers in a data matrix were attached to a dart board, and a random number of darts were thrown at the board. Each number hit by a dart would be missing. In this example, each observation is missing with probability that does not depend on the values observed. This model applies to keypunching errors, among others. The data are clearly MAR.

Missing pretest. In a study of educational achievement a teacher forgets to give one of the pretests. The missing pretest scores are missing at random.

Randomized experiment. In an experiment to compare a treatment and a control, each unit in the study is assigned at random to one group or the other. All units assigned to the treatment group are missing the score they would have obtained had they been given the control rather than the treatment. Their control score is then missing at random. In this sense, all randomized experiments have data that are missing at random, and analysis of the data ignoring the processes that caused missing data (randomization) is permissible.

Turkey tenderness. In an experiment to study a method of grading turkeys for tenderness, a sample of $n = 17$ turkeys had X = empirical estimate of tenderness measured, where X is a score from 1 (very tender) to 5 (very tough). The turkeys were then dressed, frozen, and, after a fixed time thawed, cooked, and a laboratory measure of Y = actual tenderness was obtained. However, during storage, three of the turkeys, all rated $X = 1$, were stolen, so their value of Y could not be recorded. The response Y is MAR because the process that caused the missingness may be a function of X, but it is not a function of Y, even though X and Y are presumably related.

The following are examples where MAR fails.

Censoring. Suppose that one variable in a study is a time to failure, but some units do not fail during the experimental period, so their failure time is not recorded. Missing at random is clearly inappropriate here.

Small reactions. A chemical concentration may not be measurable on some units if the actual value is less than the minimum amount measurable on the equipment being used. Again, failure to measure depends on the values that would be observed, so MAR does not hold.

Achievement test score. In a study of educational achievement, two tests are available, and teachers give the better students a harder test, and the other students an easier test. The scores the students would have gotten on the tests they did not take are not missing at random.

The following is an example for which the MAR assumption is in doubt.

Subjects dropping out of an experiment. At the beginning of an experiment, subjects are divided into several groups at random. Within each group, a different treatment is applied (the treatment might be the administration of a drug, a learning method, etc.). At the end of the experiment, perhaps several weeks later, post-treatment scores must be obtained, but some of the subjects are no longer in the study. For example, if carried out on humans, some may have quit the study or refused to continue. For these subjects, post-treatment scores are not known.

If the reason for dropping out of the study was completely unrelated to the missing variable, MAR is reasonable. For example, human subjects who drop out because they move to a different city will generally (though not always) result in data that are MAR. On the other hand, subjects who drop out because they are not doing well will usually invalidate the MAR assumption.

The importance of the distinction between MAR and non-MAR is this: *If observed data are not MAR, then any inferences that ignore the causes of the missing data may be seriously in error.* For non-MAR data, a reliable analysis will require building a model to explain the absence of the missing data, often a very difficult task, and this model must be incorporated into the data analysis. When the data are MAR, the process causing missing data can be ignored. All of the techniques and methods described here are appropriate only for data that are MAR.

Careful consideration of the process that generated the data is the best diagnostic tool for studying the assumption of MAR, as illustrated in each of the above examples.

10.2   Handling incomplete data by filling in or deleting

The simplest method for analyzing data with a few missing observations is to delete cases, variables, or a combination of the two, and obtain a data set that has no missing data. Then, usual techniques of fitting, testing, and case analysis can be used as long as the assumption of MAR is applicable to the data, since the data set obtained by deletion will be representative of the whole. Of course, if the MAR assumption is questionable, then deletion may also be questionable, since any resulting inferences may not be applicable to the entire sample or to any underlying population that the original sample is thought to represent.

If the proportion of cases with unobserved values is small, then case deletion is the most attractive method available, since the fewest assumptions are required. Certain rules about case deletion can be given. Cases missing either observed values for the response Y or for all of the independent variables should be deleted. Cases with several unobserved variables are strong candidates for deletion since they tend to contain little useful information and may introduce more variability than precision into many analyses.

Variable deletion is generally more complicated, but certain guidelines can be applied. For prediction problems, variables that are often unobserved in the data may also tend to be unobserved in the future, and are therefore poor predictors regardless of how highly correlated they are with the response. Also, in large studies, a combination of other variables may give as much information (or nearly as much) about the response as would the variable that is deleted. Here, collinearity works to the advantage of the investigator.

Fill-in methods. Filling in for the missing data achieves more or less the same outcome as deleting: a complete data set results and, with some modification, usual methods of estimation and case analysis can be applied (testing and confidence statements are not as clear). The problem is deciding on values to fill in for the missing data. Unfortunately, the correct values to fill in are unknown; if they were known, they would not be missing. The prudent approach to this problem is to use all available information to obtain plausible values for the missing data, fill in various combinations of them, and attempt to monitor the effect of the missing data on the estimation of parameters, on model building, and on case analysis.

If additional information external to the data about the missing values is available, it should be used to help choose values to fill in. For data collected in time sequence, for example, missing values can be reliably estimated from observed values for that variable immediately prior to, and after, the missing value. In the Berkeley Guidance Study (Problem 2.1), a longitudinal study of growth of children, measurements such as height and weight were obtained for each child at intervals of $\frac{1}{2}$ year up to age 18 years. If the weight of a child at age 8 years were missing, it could be estimated by averaging the observed weights of that child at age $7\frac{1}{2}$ years and $8\frac{1}{2}$ years. In other problems, similar considerations in the experiment might severely restrict the set of plausible values to some relatively small set.

Lacking additional information bearing directly on the value to be estimated, data that are fully observed can be used to obtain an estimation equation for data that are only partially observed. As a simple example, suppose that there are two predictors X_1 and X_2, both of which are occasionally missing (cases missing both X_1 and X_2 are assumed deleted from the data set). For the fully observed data suppose that the scatter plot of X_1 versus X_2 is as given in Figure 10.1. Actually, this figure is rather

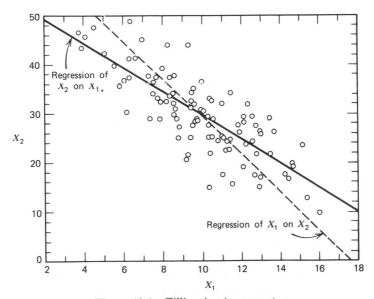

Figure 10.1 Filling in via regression

optimistic, since it was generated by making the (X_1, X_2) pairs bivariate normal. The cloud of points is more or less elliptical, and the apparent regression of X_2 on X_1 is linear. One problem of interest is to obtain values to fill in for X_2 for cases in which X_1 is observed, but X_2 is not. One could compute the regression of X_2 on X_1 for the complete data and use predictions from this fitted equation to estimate X_2 for cases with observed values of X_1 but unobserved values for X_2. This would correspond to filling in cases so that they lie along the solid line drawn on Figure 10.1, the line of the regression of X_2 on X_1. Similarly, to fill in for X_1, values filled in would fall along the dashed line, the regression of X_1 on X_2. Clearly, this process relies on the assumption that the cases to be filled in are similar to those that are complete, so a model fit to the latter gives information about the former.

Since this method tends to make the filled-in cases fall as near the center of the data swarm as possible, the influence of these cases on the resulting data analysis is made small. If several of the variables in a single case are filled in via regression, the influence of this case may be very small indeed. For example, in a four-variable problem, X_1, X_2, and X_3 could each be regressed on X_4 for the fully observed cases, and then fill-in values could be estimated for them for cases with X_4 observed, but not X_1, X_2, or X_3. If it turned out that the true values of the X_j's for this case were far from the middle of the data, and this case should be influential, then filling in via regression will lose this information.

Filling in via regression has many variations. For example, a model to predict one of the dependent variables from some of the others could be built using the methodology derived in this book. This could include subset selection, transformation of scales, and so on. Alternatively, simpler methods, such as using only one or two variable models to fill in can be used. As long as the data are a sample from a population (and the regression of the missing variable to be filled in on its predictors is linear), any of these variations may be of use.

Adjusting the analysis with filled-in values. If a data set is filled in using any ad hoc method, a usual analysis can be carried out, except that the number of degrees of freedom for error should be reduced by one for each value filled in. Naturally, this limits the number of values filled in to $n - p'$, the number of degrees of freedom for error, as is quite reasonable, since $n - p'$ values may be found to make $RSS = 0$. In addition, testing and confidence procedures that generally require distributional assumptions are not exactly correct, and only approximate statements are possible.

Depending on the nature of the data filled in, test procedures may be seriously in error. Also, any claims of optimal properties for the estimators, such as unbiasedness or minimum variance, cannot be made.

10.3 Maximum likelihood estimates assuming normality

If one can assume that all the original data (observed and unobserved) for Y and all the X's are a sample for a known multivariate distribution, then it is possible to use the assumed distribution to find maximum likelihood estimates of the parameters from the observed data. (It is interesting that finding maximum likelihood estimates can be shown to be equivalent to a fill-in method.) Computing these estimates usually requires an iterative procedure, as given by Dempster *et al.* (1977), Beale and Little (1975), Orchard and Woodbury (1972), Hartley and Hocking (1971), and Buck (1960). Rubin (1974) describes conditions under which noniterative computations are possible.

Aside from computational considerations, the main difficulty in using the maximum likelihood method is the need for distributional assumptions. Typically, multivariate normality for all the variables is used. This assumption is often unacceptable since dummy variables, variables measured on limited ranges, and variables set by the experimenter are typical in regression problems. However, putting aside these difficulties, the maximum likelihood estimates based on the incomplete data do use all of the information available in a set of data, and they should therefore be considered for use when they are relevant. Further discussion of data analysis based on maximum likelihood estimation with incomplete data is given by Little (1979).

10.4 Missing observation correlation

All of the calculations of aggregate analysis depend on only a few summary statistics, namely sample means, variances, and covariances. A commonly applied method with missing data is to compute each of these sample statistics using all the observed data, and then proceed with the least squares analysis as if these were all computed from complete data. This method is often called the missing observation correlation method.

For example, suppose a data set consists of $n = 8$ cases on four variables as shown below.

Case Number	Y	X_1	X_2	X_3
1	X	X	X	
2	X	X	X	X
3	X	X		
4	X	X		X
5	X	X		
6	X	X		X
7	X		X	X
8	X		X	

X = observed   blank = missing

Then the correlation between Y and X_1 would be computed using the first 6 cases, the correlation between Y and X_3 would be based on cases 2, 4, 6, and 7, and so on. For these data, as many as six cases will be used to estimate a correlation or as few as two are used (for the correlation between X_2 and X_3).

The success of this method depends on the assumption that the data are a sample from a population so that each correlation in the missing observation correlation matrix is an estimate of a population correlation. As long as all the correlations are small, and sample sizes are not too small, the results of this method can be expected to be reasonable. However, in the presence of large correlations, severe problems may result, including computation of negative sums of squares, or R^2 values larger than one. This happens because the computed correlation matrix may not be positive definite, as each correlation may be based on different cases. Because of this, routine use of the missing observation correlation matrix should be avoided. With it, interpretations of tests, estimates, and predictions are nearly impossible, and no meaningful case analysis has been suggested.

10.5 General recommendations

Since missing data cannot always be avoided, some general guidelines for handling a data set that is incomplete are needed. First, study the pattern of the missing data. This can often be best done using a special program, such as BMDPAM in the BMDP series (Dixon and Brown, 1979). Much can be gained by determining which variables are partially observed, which cases have many missing variables, and the overall pattern of missing data.

If the missing at random assumption is reasonable, then decisions to delete cases or variables can be made. To the data remaining after deletion, fill-in methods may be applied. Or, better yet, several methods should be applied and compared. More than one analysis (e.g., pass through a computer) should be expected.

We conclude with a warning: *Do not let a computer program decide what is to be done about missing data.* Many large-scale packages can handle missing data, often using one or more of the methods described here, or modifications of them. Keep in mind that the writer of the program did not know the details of your problem. The default analysis used in the program is almost certainly not adequate for a complete analysis.

11

NONLEAST SQUARES ESTIMATION

The most common method of fitting linear models uses the least squares estimators discussed throughout most of this book. When the errors are independent, with common variance, and perhaps are normally distributed, the use of least squares results in estimators that are well known to be the best in a class of potential estimators on a specific criterion. In this book, since the assumptions necessary are often untenable, the optimal properties of least squares have been largely neglected. Also, and more importantly, for the optimality results to apply, the linear model used should be valid and known except for the values of a few parameters. This assumes, among other things, that all variables are properly transformed, that there are no outliers, and that variable selection or other model building is not at issue. When data are used to make decisions about the model, claiming optimality for least squares estimators is impossible. Rather, least squares provides a standard methodology for computing estimates and for fitting models.

In recent years, interest in biased nonleast squares estimators has increased. The basic idea is that, in problems where the question of optimality makes sense, nonleast squares estimators may be better on some criteria than least squares estimators. For example, suppose the model is given by

$$\mathbf{Y} = \mathbf{1}\beta_0 + \mathfrak{X}\boldsymbol{\beta} + \mathbf{e} \tag{11.1}$$

where $\mathfrak{X}$ is $n \times p$ in deviations from the mean form, $\boldsymbol{\beta}$ is $p \times 1$, excluding a term for the intercept, and $\text{cov}(\mathbf{e}) = \sigma^2 \mathbf{I}$. The least squares estimator is

$\hat{\beta} = (\mathcal{X}^T \mathcal{X})^{-1} \mathcal{X}^T \mathbf{Y}$. As mentioned in Chapter 2, $\hat{\beta}$ is the minimum variance unbiased estimator, $\mathrm{var}(\hat{\beta}) = \sigma^2 (\mathcal{X}^T \mathcal{X})^{-1}$.

However, suppose that we enlarge the class of estimators considered to include biased estimators, and consider as a criterion function the sum of the mean square errors of estimating the β_j's, SMSE, where

$$SMSE = \sum_{j=1}^{p} E(\hat{\beta}_j - \beta_j)^2$$

$$= \sum_{j=1}^{p} \left\{ \mathrm{var}(\hat{\beta}_j) + \left[\mathrm{bias}(\hat{\beta}_j) \right]^2 \right\} \qquad (11.2)$$

$$= E(\hat{\beta} - \beta)^T (\hat{\beta} - \beta)$$

This is a different criterion function than the one that leads to the least squares estimator, so it should be no surprise that other estimators may be better on this criterion. Before describing such estimators, it is useful to study (11.2) carefully. The important characteristics are (1) all the β_j enter in equally, so, implicitly, interest is equal in each of them; (2) SMSE is not scale invariant, so scaling of the X's is critical; (3) interest centers on estimation of parameters rather than any other aspect of regression analysis; and (4) covariance between estimators is ignored.

To deal with the lack of invariance, each column of $\mathcal{X}$ is usually standardized so that $\mathcal{X}^T \mathcal{X}$ is the sample correlation matrix (Marquart and Snee, 1975; Obenchain, 1975). As further notation let $\lambda_1 \geqslant \lambda_2 \geqslant \cdots \geqslant \lambda_p$ be the ordered eigenvalues of $\mathcal{X}^T \mathcal{X}$. Then, Hoerl and Kennard (1970a) have noted that, for $\hat{\beta} = $ least squares estimate,

$$SMSE = E(\hat{\beta} - \beta)^T (\hat{\beta} - \beta) = \sigma^2 \, \mathrm{trace}(\mathcal{X}^T \mathcal{X})^{-1} = \sigma^2 \sum_{j=1}^{p} \lambda_j^{-1} \quad (11.3)$$

But $E(\hat{\beta} - \beta)^T (\hat{\beta} - \beta) = E(\hat{\beta}^T \hat{\beta}) - \beta^T \beta$. Substituting this into (11.3),

$$E(\hat{\beta}^T \hat{\beta}) = \beta^T \beta + \sigma^2 \sum \lambda_j^{-1}$$

$$\geqslant \beta^T \beta + \sigma^2 \lambda_p^{-1} \qquad (11.4)$$

Thus even though $\hat{\beta}$ is unbiased for β, $\hat{\beta}^T \hat{\beta}$ is not unbiased for $\beta^T \beta$ and if the smallest eigenvalue λ_p is near zero, then on the average $\hat{\beta}^T \hat{\beta}$ will be much too great (recall that λ_p near zero is symptomatic of collinearity). When λ_p is small, and (11.2) is of interest, substantial gain over least squares is possible.

Most alternative estimators to be considered here have the common characteristic that they will give an estimate $\tilde{\beta}$ of β that is shorter than least squares ($\tilde{\beta}^T \tilde{\beta} < \hat{\beta}^T \hat{\beta}$ where $\tilde{\beta}$ is the alternative estimator), so these techniques will shrink the least squares estimators (shrinkage is generally toward the origin **0**, although the choice of an origin to shrink towards can be modified). We have already encountered one of these shrinkers: Subset selection, where enough elements of $\boldsymbol{\beta}$ are set to zero to make the smallest eigenvalue used in (11.4) relatively larger, and the resulting $\hat{\boldsymbol{\beta}}^T\hat{\boldsymbol{\beta}}$ is not too much greater than $\boldsymbol{\beta}^T\boldsymbol{\beta}$.

Before turning to other specific estimators, it is useful to reexpress the original problem (11.1) in cannonical form, in which the columns of $\mathcal{X}$ are replaced by the p orthogonal variates (the principal components). From Section 7.6, there is a $p \times p$ orthogonal matrix whose columns are the eigenvectors of $\mathcal{X}^T\mathcal{X}$, say $\mathbf{U}$ ($\mathbf{UU}^T = \mathbf{U}^T\mathbf{U} = \mathbf{I}$), and a $p \times p$ diagonal matrix $\mathbf{D}$ with diagonal elements $\lambda_1 \geqslant \lambda_2 \geqslant \cdots \geqslant \lambda_p \geqslant 0$, such that

$$\mathcal{X}^T\mathcal{X} = \mathbf{UDU}^T \qquad (11.5)$$

Letting $\mathbf{Z} = \mathcal{X}\mathbf{U}$ (so the columns of $\mathbf{Z}$ are the principal components of $\mathcal{X}$) and $\boldsymbol{\alpha} = \mathbf{U}^T\boldsymbol{\beta}$, then

$$\mathbf{Y} = \mathbf{1}\beta_0 + \mathbf{X}\boldsymbol{\beta} + \mathbf{e}$$
$$= \mathbf{1}\beta_0 + \mathbf{X}(\mathbf{UU}^T)\boldsymbol{\beta} + \mathbf{e}$$
$$= \mathbf{1}\beta_0 + \mathbf{Z}\boldsymbol{\alpha} + \mathbf{e} \qquad (11.6)$$

and model (11.6) is equivalent to (11.1). Estimates of $\boldsymbol{\beta}$ and of $\boldsymbol{\alpha}$ are related by the equation

$$(\text{estimate of } \boldsymbol{\alpha}) = \mathbf{U}^T(\text{estimate of } \boldsymbol{\beta}) \qquad (11.7)$$

so computation of either $\hat{\boldsymbol{\alpha}}$ or $\hat{\boldsymbol{\beta}}$ is equivalent. However, since $\mathbf{Z}^T\mathbf{Z} = \mathbf{D}$, $\text{var}(\hat{\boldsymbol{\alpha}}) = \sigma^2(\mathbf{Z}^T\mathbf{Z})^{-1} = \sigma^2\mathbf{D}^{-1}$, the $\hat{\alpha}_j$ are independent of each other. Also, $\text{var}(\hat{\alpha}_p) = \sigma^2\lambda_p^{-1}$, and thus $\hat{\alpha}_p$ has larger variance than any other possible estimate—the information in the data concerning the pth column of $\mathbf{Z}$ is less than the information for any other possible variable that is a combination of the original X's.

The methods considered in Sections 11.1 to 11.5 have the effect of shrinking the $\hat{\alpha}_j$ in a way that is easy to describe. The effect of shrinking on the $\hat{\beta}_j$ is transmitted via (11.7), and may be more complicated. A recent survey of shrunken estimators is given by Draper and Van Nostrand (1979).

11.1   Ridge regression

The ridge regression estimator $\tilde{\boldsymbol{\beta}}(RR)$, is defined by Hoerl and Kennard (1970a; 1970b), for some $k \geqslant 0$, by

$$\tilde{\boldsymbol{\beta}}(RR) = (\mathcal{X}^T\mathcal{X} + k\mathbf{I})^{-1}\mathcal{X}^T\mathbf{Y} \qquad (11.8)$$

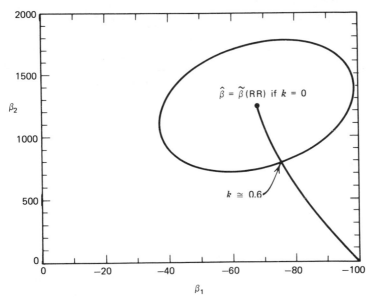

Figure 11.1   Path of the ridge estimator (based on Figure 4.3).

If $k = 0$, $\tilde{\beta}(\mathrm{RR}) = \hat{\beta}$, the least squares estimator, while larger k will move $\tilde{\beta}(\mathrm{RR})$ away from least squares, and increase the bias in the estimate. The ridge parameter k indexes an infinity of possible ridge estimators.

Figure 11.1 illustrates the effect of increasing k on the ridge estimator. The data used to generate this figure were also used in Figure 4.3; the ellipse is the 95% confidence ellipsoid for ($\hat{\beta}_1$, $\hat{\beta}_2$) based on least squares. The ridge estimates were computed from (11.8), and the curved line gives a plot of $\tilde{\beta}(\mathrm{RR})$ as k increases from 0. At $k = 0$, $\tilde{\beta}(\mathrm{RR}) = \hat{\beta}$. When $k \cong 0.6$, $\tilde{\beta}(\mathrm{RR})$ falls on the boundary of the 95% confidence ellipsoid. For any $k < 0.6$, $\tilde{\beta}(\mathrm{RR})$ is inside the ellipsoid, while larger values put $\tilde{\beta}(\mathrm{RR})$ outside the ellipsoid.

Generally, k is an unknown parameter that can be set by the analyst. Large values of k correspond to increased bias but lower variance, so a value of k must be chosen to balance bias against variance. The basic benefit of ridge regression is summarized by the following result of Hoerl and Kennard (1970a): For every fixed $\mathscr{X}$ and β in the model (11.1) there is a k_0 such that for all $0 < k < k_0$, the SMSE of $\tilde{\beta}(\mathrm{RR})$ is less than the SMSE of $\hat{\beta}$. However, Thisted (1978b) has shown that for any fixed $k > 0$ and any $\mathscr{X}$, there is a regression problem (i.e., an actual value β) such that the SMSE of $\hat{\beta}$ is less than the SMSE of $\tilde{\beta}(\mathrm{RR})$. Thus if the object of using ridge regression is to minimize SMSE, k must be estimated from the data.

Methods of estimating k have been suggested by many writers, including Hoerl and Kennard (1970a); a survey of suggested methods is given by Draper and Van Nostrand (1978).

Cannonical form. In the cannonical form, the ridge estimate $\tilde{\alpha}_j(\text{RR})$ of α_j can be shown to be equal to

$$\tilde{\alpha}_j(\text{RR}) = \frac{\lambda_j}{\lambda_j + k} \hat{\alpha}_j \quad j = 1, 2, \ldots, p \tag{11.9}$$

where $\hat{\alpha}_j$ is the least squares estimator. The effect of ridge regression is to leave least squares estimates nearly unchanged if λ_j is much larger than k, but if λ_j is small compared to k, then the corresponding $\tilde{\alpha}_j(\text{RR})$ will be much smaller than $\hat{\alpha}_j$. Thus information coming from highly variable estimates is downweighted.

Relationship to Bayes rules. Prior information about parameters is formally incorporated into a problem through the use of Bayesian methods. If we assume that the vector $\boldsymbol{\alpha}$ from (11.6) is drawn from a normal distribution

$$\boldsymbol{\alpha} \sim N(\mathbf{0}, k^{-1}\mathbf{I}) \tag{11.10}$$

and if we also assume that $\mathbf{e} \sim N(\mathbf{0}, \sigma^2\mathbf{I})$, then the Bayes estimate of $\boldsymbol{\beta}$ in (11.1) is given by (11.8). The value k^{-1}, the variance of each α_j, represents the prior variability in $\boldsymbol{\alpha}$, and, if the assumptions made are reasonable, the estimator (11.8) is very attractive. However, an assumption of $\mathbf{0}$ prior mean is rarely reasonable (of course methods can be modified for other prior means); if it were tenable, collection and analysis of data would probably never have been done.

Examples of the use of ridge regression are given by Marquart and Snee (1975); also, see Smith and Campbell (1980).

11.2   Generalized ridge regression

One generalization of the ridge regression rule is to replace the ridge parameter k by a vector of parameters $(k_1, k_2, \ldots, k_p)$, so that in the cannonical formulation, the estimates of α_j the generalized ridge estimate, $\tilde{\alpha}_j(\text{GR})$, are

$$\tilde{\alpha}_j(\text{GR}) = \frac{\lambda_j}{\lambda_j + k_j} \hat{\alpha}_j \quad j = 1, 2, \ldots, p \tag{11.11}$$

Generalized ridge regression rules therefore apply different ridge parameters to each λ_j, allowing for many different patterns of shrinking the estimates. In terms of the original coordinates, define $\mathbf{G}$ to be the $p \times p$ matrix such that, if $\mathbf{K}$ is the diagonal matrix with $k_1, k_2, \ldots, k_p$ on the diagonal, $\mathbf{G} = \mathbf{U}\mathbf{K}\mathbf{U}^T$, with $\mathbf{U}$ as defined near (11.6). Then, the generalized ridge estimator is (Hoerl and Kennard, 1970a; Bingham and Larntz, 1977)

$$\tilde{\beta}(\text{GR}) = (\mathfrak{X}^T\mathfrak{X} + \mathbf{G})^{-1}\mathfrak{X}^T\mathbf{Y} \qquad (11.12)$$

Note in particular if $k_1 = k_2 = \cdots = k_p = k$, then $\mathbf{K} = k\mathbf{I}$, and $\mathbf{G} = \mathbf{U}(k\mathbf{I})\mathbf{U}^T = k\mathbf{I}$, so ridge regression is a special case of generalized ridge regression.

Various methods of estimating or otherwise determining $k_1, \ldots, k_p$ have been suggested; when a criterion like (11.2) is of interest, methods suggested by Strawderman (1978), Berger (1975), and Thisted (1978a) are relevant. Other criteria will lead to alternative choices for the k_j's, one of which is considered in the next section.

11.3   Regression on principal components

One special case of generalized ridge regression results in performing the regression on some of the principal component vectors. If each k_j is set either to 0 or allowed to approach $+\infty$, then the regression on principal component estimators $\tilde{\alpha}_j(\text{PC})$, $j = 1, 2, \ldots, p$ are given by

$$\tilde{\alpha}_j(\text{PC}) = \begin{cases} \hat{\alpha}_j & \text{if} \quad k_j = 0 \\ 0 & \text{if} \quad k_j \to +\infty \end{cases} \qquad (11.13)$$

The corresponding $\tilde{\beta}(\text{PC})$ is found by substituting $\tilde{\alpha}(\text{PC})$ into (11.7) (Marquart, 1970; Mansfield et al., 1977). If the k_j's for small λ_j are allowed to approach $+\infty$, the regression on principal components can have much smaller SMSE than least squares (Dempster et al., 1977). A computer program for regression on principal components is included in the BMDP series (BMDP4R, Dixon and Brown, 1979).

11.4   James-Stein estimators

These estimators are based on the result of Stein (1956) and James and Stein (1961), that in the problem of estimating the mean of three or more normal distributions, the sample average vector is inadmissable—that is, there exist estimators that are always better (in some sense) than the sample averages. These improved estimators are shrunken estimators in which the vector of sample means is shrunken toward zero or some other point or subspace; see Efron and Morris (1973; 1975) for more details.

The usual James-Stein estimators can be obtained in a Bayesian framework by assuming that each of the α_j's is independent and normally distributed, with mean zero and variance proportional to σ^2/λ_j. (To obtain ridge estimators, one assumes variance proportional to σ^2 (Dempster, 1973; Goldstein and Smith, 1974; Sclove, 1968; Rolph, 1976)). Then, the James-Stein estimator $\tilde{\beta}(JS)$, is given by

$$\tilde{\beta}(JS) = (1 - \hat{B})\hat{\beta} \qquad (11.14)$$

A common choice of $\hat{B}$ is

$$\hat{B} = \text{minimum}\left\{1, \frac{(p-2)(n-p')\hat{\sigma}^2}{\hat{\beta}^T\hat{\beta}(n-p'+2)}\right\}$$

This form of the James-Stein rule (there are others) has the undesirable property of proportionally shrinking each estimator. This rule can also be written

$$\hat{\beta}(JS) = \left[\mathfrak{X}^T\mathfrak{X} + \hat{B}(1-\hat{B})^{-1}\mathfrak{X}^T\mathfrak{X}\right]^{-1}\mathfrak{X}^T Y \qquad (11.15)$$

for comparison to other estimators in this chapter. The class of proportionally shrinking James-Stein rules is generated by modifying the definition of $\hat{B}$.

11.5   Summary of shrunken estimators

All of the shrunken estimators can result in improved SMSE in some circumstances. Draper and Van Nostrand (1978) report that the improvement over least squares will be very small whenever the parameter β is well estimated—that is, if collinearity is not a problem, and β is not too close to zero. On the other hand, if β is poorly estimated, either because of collinearity or β being near zero, the shrunken estimates may provide a substantial improvement over least squares. But the importance of this improvement is far from clear. If β is near zero, the independent variables are only slightly related to the response, and more precise estimation may not be of any value. If the data are collinear, then for some combinations of parameters, the data contain little reliable information to be used for estimation. While least squares result in poor estimates for these combinations, the shrunken methods produce relatively more precise estimates of poorly determined quantities.

11.6   Robust regression

As we have seen in earlier chapters, the least squares estimator of a parameter vector can be very sensitive to outliers or to cases with large v_{ii}. Thus naive application of least squares to obtain estimates is not a robust procedure, in that modest changes in the data set can result in substantial changes in the estimates and in the conclusions to be drawn from the analysis. However, if case analysis is added to the analysis of a regression problem, then the entire procedure of model fitting is robust, since potentially influential cases will be found, and their effects studied. As a result, use of the methodology described in this book, which includes using least squares to obtain estimates, is a robust procedure.

Some investigators use alternative estimation methods that are less sensitive to changes in the data. Generally called robust regression methods, these are designed to be used when errors are drawn from a distribution that is symmetric, but has longer tails than the normal distribution (that is, more large deviations are observed than would be expected if the errors were normally distributed). As with the least squares estimators, use of robust methods assumes that the functional form fit for the model is in fact the correct one, and estimates are robust against long-tailed error distributions, but not necessarily against incorrect specification of a model.

One important class of robust estimates are called *M-estimates*, a shorthand for maximum likelihood type estimates. These are obtained by choosing estimates to minimize a function of the residuals other than their sum of squares. Let $\tilde{\beta}$ be the estimator of β, y_i the ith observed value for the response, and x_i^T the ith row of the matrix X. Then, for some function ρ (choice of ρ determines the estimator), $\tilde{\beta}$ is chosen to minimize

$$\sum_{i=1}^{n} \rho(y_i - x_i^T\tilde{\beta})   \tag{11.16}$$

By varying the choice of the function ρ, the character of the resulting estimator can be changed. For example, $\rho(z) = z^2$ corresponds to the usual least squares estimator. Another popular choice is $\rho(z) = |z|^f$ where f is often taken to be a positive constant smaller than 2. The choice of $f = 1$ gives the least absolute deviations estimator, while smaller values for f will tend to give increasingly smaller weight to large residuals. Another important choice is

$$\rho(z) = \begin{cases} z^2/2 & \text{if } |z| \leqslant c \quad \text{for } c \text{ a fixed positive constant} \\ c|z| - c^2/2 & |z| > c \end{cases}$$

This choice for ρ also results in downweighting cases with large residuals relative to the weight of cases with small residuals. Computation of most M-estimates requires an iterative procedure.

The literature on robust methodology has become quite large in the last few years. Some important references include Andrews et al. (1972), Andrews (1974), Huber (1973; 1977), and Mosteller and Tukey (1977).

APPENDIX

1A.1 A formal development of the simple regression model

Suppose we have two quantities, an independent variable X and a response Y, and the actual relationship between X and Y is determined by some unknown function f, so that

$$Y = f(X) \qquad (1A.1)$$

By collecting data, we want to study f, and thus study the relationship between X and Y. To this end, for each of n units or cases, we observe values x_i of X and y_i of Y, $i = 1, 2, \ldots, n$, where

$$y_i = f(x_i) + \epsilon_i \qquad (1A.2)$$

and ϵ_i is a random error representing variability in the observational process due to measurement error, neglected factors, and the like.

Now suppose that the shape of the unknown f can be approximated by a straight line. For this to be valid, the scales of X and/or Y may need to be changed or else the range of values of X considered may be limited. In any case, view the straight line as a first try at approximating f, and, if later analysis suggests this model to be inadequate, other analyses should be substituted. Thus $f(x)$ is approximated by $\beta_0 + \beta_1 x_i$ for some β_0, β_1 and

$$f(x_i) = \beta_0 + \beta_1 x_i + \delta_i \qquad (1A.3)$$

where δ_i is the fixed or lack of fit error reflecting the inadequacy of the straight line in matching f, $\delta_i = f(x_i) - \beta_0 - \beta_1 x_i$. For the simple regression model to be useful, the δ_i should be small (negligible) when compared to

239

the ϵ_i. Combining (1A.2) and (1A.3), and defining $e_i = \epsilon_i + \delta_i$, we get the simple regression model

$$y_i = \beta_0 + \beta_1 x_i + e_i \qquad i = 1, 2, \ldots, n \qquad (1A.4)$$

where the e_i's, as in the text, consist of a fixed component and a random component.

In this development, we have taken the x_i's to be measured without error. Including errors in the X's complicates some of the analysis, and, whenever possible, it is useful to assume that errors in X are relatively small. Methodology for checking this assumption is a topic in Chapter 3.

1A.2   Means and variances of random variables

Suppose we let $u_1, u_2, \ldots, u_n$ be random variables, and also let a_0, $a_1, \ldots, a_n$ be $n + 1$ known constants.

E **notation.** The symbol $E(u_i)$ is read as the expected value of the random variable u_i. The term "expected value" is the same as the term "mean value," or less formally, the arithmetic average for a very large sample size. The statement $E(u_i) = 0$ means that the average value we would get for u_i, if we sampled its distribution repeatedly, is 0; however, any specific realization of u_i that we observe is likely to be nonzero.

The expectation of a sum of random variables may be symbolically expressed by the following two equations:

$$E(a_0 + a_1 u_1) = a_0 + a_1 E(u_1) \qquad (1A.5)$$

$$E\left(a_0 + \sum a_i u_i\right) = a_0 + \sum a_i E(u_i) \qquad (1A.6)$$

For example, suppose that $u_1, u_2, \ldots, u_n$ make up a random sample, and $E(u_i) = \mu$, a constant value for all $i = 1, 2, \ldots, n$. Then, the expected value of the sample average of the u_i's, $\bar{u} = \sum u_i / n = (1/n)u_1 + (1/n)u_2 + \cdots + (1/n)u_n$, can be found from equation (1A.6) with $a_i = 1/n$, $i = 1, 2, \ldots, n$, and $a_0 = 0$. Thus

$$E(\bar{u}) = \sum \left(\frac{1}{n}\right) E(u_i) = \left(\frac{1}{n}\right) \sum \mu = \left(\frac{1}{n}\right) n\mu = \mu \qquad (1A.7)$$

so that the sample average is an unbiased estimate of the mean μ.

var Notation. The symbol var(u_i) is read as the variance of u_i. The variance is defined by the equation var(u_i) = $E[u_i - E(u_i)]^2$ = the expected square of the difference between an observed value for u_i and its average value. The larger var(u_i) is, the more variable observed values for u_i are likely to be. The symbol σ^2 is often used for a variance, or σ_u^2 might be used for the variance of the u's when several variances are being discussed.

The general rule for the variance of a sum of random variables (if the variables are uncorrelated) is

$$\text{var}\left(a_0 + \sum a_i u_i\right) = \sum a_i^2 \text{var}(u_i) \qquad (1A.8)$$

The a_0 term vanishes: the variance of a constant is zero. We can now apply this equation to find the variance of the sample average, assuming that the u_i's are uncorrelated with common variance var(u_i) $= \sigma_u^2$:

$$\text{var}\left(\sum\left(\frac{1}{n}\right)u_i\right) = \sum\left(\frac{1}{n}\right)^2 \text{var}(u_i) = n\left(\frac{1}{n}\right)^2 \sigma_u^2 = \frac{\sigma_u^2}{n}$$

cov Notation. The symbol cov(u_i, u_j) is read to be the covariance between the random variables u_i and u_j and is defined by the equation cov(u_i, u_j) $= E(u_i - E(u_i))(u_j - E(u_j))$. The covariance describes the way two random variables jointly vary. If the two variables are independent, then they are uncorrelated (but not necessarily conversely). If $i = j$ in the definition, then we see that cov(u_i, u_i) $=$ var(u_i) by the definition of the latter symbol. The rule for covariance is

$$\text{cov}(a_0 + a_1 u_1, a_3 + a_2 u_2) = a_1 a_2 \text{cov}(u_1, u_2) \qquad (1A.9)$$

Often, a scale-free version of the covariance is used in its place. This is called the *correlation coefficient*, abbreviated as corr(u_i, u_j), and is defined by the equation

$$\text{corr}(u_i, u_j) = \frac{\text{cov}(u_i, u_j)}{\sqrt{\text{var}(u_i)\text{var}(u_j)}} \qquad (1A.10)$$

The correlation does not depend on the units for the random variables and has a value between $+1$ and -1. If the correlation is zero, then the variables u_i and u_j are uncorrelated; this will happen only if cov(u_i, u_j) $= 0$.

The general form of the variance of linear combination of random variables depends both on the variances of the variables, and on their covariances, according to the following rule:

$$\text{var}(a_0 + \sum a_i u_i) = \sum_{i=1}^{n} a_i^2 \text{var}(u_i) + 2\sum_{i=1}^{n-1}\sum_{j=i+1}^{n} a_i a_j \text{cov}(u_i, u_j) \quad (1A.11)$$

1A.3   Least squares

The least squares estimates of the parameters β_0 and β_1 in the simple regression model are found by minimizing the residual sum of squares

$$RSS = \sum_{i=1}^{n} (y_i - \hat{\beta}_0 - \hat{\beta}_1 x_i)^2 \tag{1A.12}$$

One method for carrying out this minimization is to differentiate (1A.12) with respect to $\hat{\beta}_0$ and $\hat{\beta}_1$, set the derivatives to zero, and solve the resulting equations for the estimates. Carrying out this plan,

$$\frac{\partial RSS}{\partial \hat{\beta}_0} = -2 \sum (y_i - \hat{\beta}_0 - \hat{\beta}_1 x_i) = 0$$

$$\frac{\partial RSS}{\partial \hat{\beta}_1} = -2 \sum x_i (y_i - \hat{\beta}_0 - \hat{\beta}_1 x_i) = 0 \tag{1A.13}$$

Upon rearranging terms (1A.13) becomes

$$\hat{\beta}_0 n + \hat{\beta}_1 \sum x_i = \sum y_i$$

$$\hat{\beta}_0 \sum x_i + \hat{\beta}_1 \sum x_i^2 = \sum x_i y_i \tag{1A.14}$$

Equations (1A.14) are called the *normal equations* for model (1.2). The data are used only through the aggregates $\sum x_i$, $\sum x_i^2$, $\sum y_i$, and $\sum y_i^2$ (or equivalently, through $\bar{x}$, $\bar{y}$, SXX, and SXY). Any two data sets with these quantities the same will have the same estimates $\hat{\beta}_0$ and $\hat{\beta}_1$. The estimates are now found by solving the normal equations (two linear equations in two unknowns):

$$\hat{\beta}_0 = \bar{y} - \hat{\beta}_1 \bar{x}$$

$$\hat{\beta}_1 = \frac{SXY}{SXX} \tag{1A.15}$$

1A.4   Means and variances of least squares estimates

The least squares estimates are linear functions of the y_i, $i = 1, \ldots, n$, and, since the y_i's are linear functions of the e_i's, we will be able to apply the results of section 1A.2 to the estimates found in 1A.3 to get the means, variances, and covariances of the estimates. In particular, assume the simple regression model

$$y_i = \beta_0 + \beta_1 x_i + e_i \qquad i = 1, 2, \ldots, n$$

is correct. By results (1A.6) and (1A.8), $E(y_i) = \beta_0 + \beta_1 x_i$, and var$(y_i) =$ var$(e_i) = \sigma^2$. Now, consider the estimate $\hat{\beta}_1$, given in (1A.15). Suppose we define the constants $c_1, c_2, \ldots, c_n$ by the equation (for each i)

$$c_i = \frac{(x_i - \bar{x})}{SXX} \qquad i = 1, 2, \ldots, n$$

Since the x_i's are considered to be fixed numbers, so are the c_i. The estimator $\hat{\beta}_1$ is equal to $\sum c_i y_i$, a linear combination of the y_i's. The mean of $\hat{\beta}_1$ is then found as

$$E(\hat{\beta}_1) = E\left(\sum c_i y_i\right) = \sum c_i E(y_i)$$

$$= \sum c_i(\beta_0 + \beta_1 x_i) = \beta_0 \sum c_i + \beta_1 \sum c_i x_i$$

But one can show by direct summation $\sum c_i = 0$, $\sum c_i x_i = 1$, giving

$$E(\hat{\beta}_1) = \beta_1 \qquad (1A.16)$$

which shows that, as long as $E(y_i) = \beta_0 + \beta_1 x_i$, $\hat{\beta}_1$ is an unbiased estimate of β_1 (also, one can easily show that $E(\hat{\beta}_0) = \beta_0$).

The variance of $\hat{\beta}_1$ is found by

$$\text{var}(\hat{\beta}_1) = \text{var} \sum (c_i y_i)$$

$$= \sum c_i^2 \, \text{var}(y_i) + 2 \sum_{i=1}^{n-1} \sum_{j=i+1}^{n} c_i c_j \, \text{cov}(y_i, y_j)$$

but cov$(y_j, y_i) = \text{cov}(\beta_0 + \beta_1 x_j + e_j, \beta_0 + \beta_1 x_i + e_i) = \text{cov}(e_j, e_i) = 0$ by assumption. Also var$(y_i) = \text{var}(e_i) = \sigma^2$ by assumption. Hence

$$\text{var}(\hat{\beta}_1) = \sigma^2 \sum c_i^2$$

but $\sum c_i^2 = 1/SXX$, so that

$$\text{var}(\hat{\beta}_1) = \sigma^2 \frac{1}{SXX} \qquad (1A.17)$$

To find the variance of $\hat{\beta}_0$, write

$$\text{var}(\hat{\beta}_0) = \text{var}(\bar{y} - \hat{\beta}_1 \bar{x})$$

$$= \text{var}(\bar{y}) + \bar{x}^2 \, \text{var}(\hat{\beta}_1) - 2\bar{x} \, \text{cov}(\bar{y}, \hat{\beta}_1)$$

Now, var$(\bar{y}) = \sigma^2/n$, var$(\hat{\beta}_1)$ is given by (1A.17), and cov$(\bar{y}, \hat{\beta}_1) = 0$. This last result can be shown by application of the rules in the last section, but it is intuitively clear because the average value $\bar{y}$ should not depend in any way on the fitted slope $\hat{\beta}_1$. Thus we get

$$\text{var}(\hat{\beta}_0) = \sigma^2\left(\frac{1}{n} + \frac{\bar{x}^2}{SXX}\right) \qquad (1A.18)$$

Finally,

$$\text{cov}(\hat{\beta}_0, \hat{\beta}_1) = \text{cov}(\bar{y} - \hat{\beta}_1\bar{x}, \hat{\beta}_1)$$

$$= \text{cov}(\bar{y}, \hat{\beta}_1) - \bar{x}\,\text{cov}(\hat{\beta}_1, \hat{\beta}_1)$$

$$= -\frac{\sigma^2\bar{x}}{SXX} \tag{1A.19}$$

1A.5   An example of round-off error

Consider the $n = 3$ numbers $x_1 = 12541$, $x_2 = 12537$, and $x_3 = 12548$. Using the uncorrected formula for SXX, but rounding the result of each multiplication to seven digits *before* adding, as is done on some computers, will give

$$\sum x_i^2 = (12541)^2 + (12537)^2 + (12548)^2$$

$$= 157276700 + 157176400 + 157452300$$

$$= 471905400$$

Now

$$\bar{x} = (12541 + 12537 + 12548)/3 = 12542$$

so

$$n\bar{x}^2 = 3(12542)^2 = 3(157301800) = 471905400$$

which gives

$$SXX = \sum x_i^2 - n\bar{x}^2 = 0.$$

The correct computation gives

$$SXX = (12541 - 12542)^2 + (12537 - 12542)^2 + (12548 - 12542)^2 = 62$$

The uncorrected formula with seven digits of accuracy gives no accurate digits in the computed solution.

2A.1   A brief introduction to matrices and vectors

A complete presentation of matrices and vectors is not attempted in this book. Two useful references on linear algebra with applications in statistics are Graybill (1969) and Searle (1961), although the material necessary for this book should be contained in any good linear algebra book.

A *matrix* (plural: matrices) is a rectangular array of numbers. We would say that $\mathbf{X}$ is an $r \times c$ matrix if it is an array of numbers with r rows and c

columns. A specific 4×3 matrix $\mathbf{X}$ is

$$\mathbf{X} = \begin{bmatrix} 1 & 2 & 1 \\ 1 & 1 & 5 \\ 1 & 3 & 4 \\ 1 & 4 & 6 \end{bmatrix} = \begin{bmatrix} x_{11} & x_{12} & x_{13} \\ x_{21} & x_{22} & x_{23} \\ x_{31} & x_{32} & x_{33} \\ x_{41} & x_{42} & x_{43} \end{bmatrix} = (x_{ij})$$

A specific element of a matrix $\mathbf{X}$ is given by x_{ij}, meaning the number in the ith row and the jth column of $\mathbf{X}$. For example, in the above matrix, $x_{23} = 5$. The usual convention in this book is to name matrices with bold-face letters and elements of a matrix by lower-case subscripted letters.

A *vector* is a matrix with 1 column. A specific 4×1 matrix $\mathbf{Y}$ (a vector of length 4) is given by

$$\mathbf{Y} = \begin{bmatrix} y_1 \\ y_2 \\ y_3 \\ y_4 \end{bmatrix} = \begin{bmatrix} 2 \\ 3 \\ -2 \\ 0 \end{bmatrix}$$

Vectors are also denoted in boldface type, and the elements of a vector are singly subscripted. Thus $y_3 = -2$ is the third element of $\mathbf{Y}$. We complicate the notation somewhat in the case of a special vector, namely, the parameter vector $\boldsymbol{\beta}$ defined by

$$\boldsymbol{\beta} = \begin{bmatrix} \beta_0 \\ \beta_1 \\ \vdots \\ \beta_p \end{bmatrix}$$

where $\boldsymbol{\beta}$ usually starts with β_0 rather than with β_1 and $\boldsymbol{\beta}$ is $(p + 1) \times 1$ if the intercept is in the model.

A *row vector* is a matrix with one row. By convention, however, only column vectors are used. If a vector is needed to represent a row, then the transpose of a column vector will be written (see below).

A *square matrix* has p rows and p columns. In many statistical applications, the square matrices that are used will be *symmetric*. We say that $\mathbf{X}$ is a symmetric $p \times p$ matrix if $x_{ij} = x_{ji}$. A very special symmetric matrix will have all elements off the main diagonal equal to zero: $x_{ij} = x_{ji} = 0$ if $i \neq j$. This is called a *diagonal* matrix. The matrices $\mathbf{C}$ and $\mathbf{D}$ below are, respectively, symmetric and diagonal:

$$\mathbf{C} = \begin{bmatrix} 7 & 3 & 2 & 1 \\ 3 & 4 & 1 & -1 \\ 2 & 1 & 6 & 3 \\ 1 & -1 & 3 & 8 \end{bmatrix} \qquad \mathbf{D} = \begin{bmatrix} 7 & 0 & 0 & 0 \\ 0 & 4 & 0 & 0 \\ 0 & 0 & 6 & 0 \\ 0 & 0 & 0 & 8 \end{bmatrix}$$

The diagonal matrix with all elements on the diagonal equal to one is called the *identity matrix*, for which the symbol $\mathbf{I}$ is used. Sometimes the identity will be written as $\mathbf{I}_n$, indicating that the identity matrix is $n \times n$:

$$\mathbf{I}_4 = \begin{bmatrix} 1 & 0 & 0 & 0 \\ 0 & 1 & 0 & 0 \\ 0 & 0 & 1 & 0 \\ 0 & 0 & 0 & 1 \end{bmatrix}$$

A *scalar* is a 1×1 matrix, an ordinary number. Scalars are usually not subscripted.

Operating with matrices: addition and subtraction. Two matrices can be added or subtracted only if they are the same size (only if they both have the same number of rows and columns). If $\mathbf{A}$ and $\mathbf{B}$ are both $n \times p$ matrices, then their sum $\mathbf{C} = \mathbf{A} + \mathbf{B}$ is also $n \times p$. The addition operation is done elementwise:

$$\mathbf{C} = \mathbf{A} + \mathbf{B} = \begin{bmatrix} a_{11} & a_{12} \\ a_{21} & a_{22} \\ a_{31} & a_{32} \end{bmatrix} + \begin{bmatrix} b_{11} & b_{12} \\ b_{21} & b_{22} \\ b_{31} & b_{32} \end{bmatrix} = \begin{bmatrix} a_{11} + b_{11} & a_{12} + b_{12} \\ a_{21} + b_{21} & a_{22} + b_{22} \\ a_{31} + b_{31} & a_{32} + b_{32} \end{bmatrix}$$

Subtraction works the same way with $(+)$ signs changed to $(-)$ signs. The usual rules for addition of numbers apply to addition of matrices, namely, commutativity $(\mathbf{A} + \mathbf{B} = \mathbf{B} + \mathbf{A})$ and associativity—$(\mathbf{A} + \mathbf{B}) + \mathbf{C} = \mathbf{A} + (\mathbf{B} + \mathbf{C}) = (\mathbf{A} + \mathbf{C}) + \mathbf{B}$.

Multiplication by a scalar. Suppose that k is a real number (i.e., a scalar). If $\mathbf{A}$ is an $r \times c$ matrix with elements (a_{ij}), then $k\mathbf{A}$ is an $r \times c$ matrix with elements (ka_{ij}); that is, each element of $\mathbf{A}$ is multiplied by k. A common use of this notation is in specifying the variance covariance matrix of a random vector. In Section 2.2, $\mathrm{var}(\mathbf{e}) = \sigma^2 \mathbf{I}_n$, which means that the variance covariance matrix is obtained by multiplying the identity elementwise by σ^2: the diagonal entries are therefore $\sigma^2(1) = \sigma^2$, and the off-diagonal entries are $\sigma^2(0) = 0$. More generally, the variance covariance matrix of a $p \times 1$ vector $\mathbf{z}$ is usually written as $\sigma_z^2 \mathbf{\Sigma}$, and, if s_{ij} is the (i, j) element of $\mathbf{\Sigma}$, then the covariance between the ith and jth element of $\mathbf{z}$ is $\sigma_z^2 s_{ij}$.

Multiplication of two matrices. Multiplication of matrices follows rules that are more complicated than are the rules for addition and subtraction. For two matrices to be multiplied together in the order $\mathbf{AB}$, the number of columns (the second dimension) of $\mathbf{A}$ must equal the number of rows (the first dimension) in $\mathbf{B}$. For example, if $\mathbf{A}$ is $n \times p$, and $\mathbf{B}$ is $p \times q$, then we can form the product $\mathbf{C} = \mathbf{AB}$, which will be an $n \times q$ matrix. If the elements of $\mathbf{A}$ are a_{ij} and those of $\mathbf{B}$ are b_{ij}, then the elements of $\mathbf{C}$ are given by c_{ij}, where the formula for c_{ij} is

$$c_{ij} = \sum_{k=1}^{p} a_{ik} b_{kj}$$

In words, this formula says that c_{ij} is formed by taking the ith row of **A** and the jth column of **B**, multiplying the first element of the specified row in **A** by the first element in the specified column in **B**, multiplying second elements, and so on, and then adding the products together.

Possibly the simplest case of multiplying two matrices **A** and **B** together occurs when **A** is $1 \times p$ and **B** is $p \times 1$. The resulting matrix will be 1×1, a scalar or ordinary number. For example, if **A** and **B** are

$$\mathbf{A} = (\ 1\ \ 3\ \ 2\ \ -1) \qquad \mathbf{B} = \begin{pmatrix} 2 \\ 1 \\ -2 \\ 4 \end{pmatrix}$$

then the product **AB** is

$$\mathbf{AB} = 1(2) + 3(1) + 2(-2) + -1(4) = -3.$$

It is not true that **AB** is the same as **BA**. In fact, for the above matrices, the product **BA** will be a 4×4 matrix:

$$\mathbf{BA} = \begin{pmatrix} 2 & 6 & 4 & -2 \\ 1 & 3 & 2 & -1 \\ -2 & -6 & -4 & 2 \\ 1 & 3 & 2 & -1 \end{pmatrix}$$

In many instances, the matrix **AB** will be well defined, but the product **BA** will not.

More generally, consider a small example. Symbolically, a 3×2 matrix **A** times a 2×2 matrix **B** is given as

$$\begin{pmatrix} a_{11} & a_{12} \\ a_{21} & a_{22} \\ a_{31} & a_{32} \end{pmatrix} \begin{pmatrix} b_{11} & b_{12} \\ b_{21} & b_{22} \end{pmatrix} = \begin{pmatrix} a_{11}b_{11} + a_{12}b_{21} & a_{11}b_{12} + a_{12}b_{22} \\ a_{21}b_{11} + a_{22}b_{21} & a_{21}b_{12} + a_{22}b_{22} \\ a_{31}b_{11} + a_{32}b_{21} & a_{31}b_{12} + a_{32}b_{22} \end{pmatrix}$$

Using numbers, an example of a multiplication of two matrices is

$$\begin{pmatrix} 3 & 1 \\ -1 & 0 \\ 2 & 2 \end{pmatrix} \begin{pmatrix} 5 & 1 \\ 0 & 4 \end{pmatrix} = \begin{pmatrix} 15 + 0 & 3 + 4 \\ -5 + 0 & -1 + 0 \\ 10 + 0 & 2 + 8 \end{pmatrix} = \begin{pmatrix} 15 & 7 \\ -5 & -1 \\ 10 & 10 \end{pmatrix}$$

In the above example, not only is $\mathbf{AB} \neq \mathbf{BA}$, but, for the matrices given, **BA** does not exist. However, the associative law holds: $\mathbf{A(BC)} = \mathbf{(AB)C}$.

Transpose of a matrix. The transpose of a matrix **X** is that matrix called $\mathbf{X}^T$ such that if x_{ij} are the elements of **X** and x'_{ij} are the elements of $\mathbf{X}^T$, then $x_{ij} = x'_{ji}$. $\mathbf{X}^T$ is obtained from $\mathbf{X}$ by taking $\mathbf{X}$ and laying it on its side. For $\mathbf{X}$ previously given,

$$\mathbf{X}^T = \begin{bmatrix} 1 & 1 & 1 & 1 \\ 2 & 1 & 3 & 4 \\ 1 & 5 & 4 & 6 \end{bmatrix}$$

The transpose of a column vector is a row vector. The transpose of a product $(\mathbf{AB})^T$ is the product of the transposes, in *opposite order*, $(\mathbf{AB})^T = \mathbf{B}^T \mathbf{A}^T$.

Suppose that $\mathbf{A}$ is an $n \times 1$ vector with elements $a_1, a_2, \ldots, a_n$. Then the product $\mathbf{A}^T \mathbf{A}$ is well defined, and it is the 1×1 matrix given by

$$\mathbf{A}^T \mathbf{A} = a_1^2 + a_2^2 + \cdots + a_n^2 = \sum a_i^2, \tag{2A.1}$$

that is, it is the sum of squares of the elements of the vector $\mathbf{A}$. The square root of this quantity is called the *norm* or the *length* of the vector $\mathbf{A}$. For example, let $\mathbf{Y}$ be the observed $n \times 1$ data vector, $\hat{\mathbf{Y}}$ the $n \times 1$ vector of fitted values. Then the residual vector is given by $\hat{\mathbf{e}} = \mathbf{Y} - \hat{\mathbf{Y}}$, and the residual sum of squares is simply $\hat{\mathbf{e}}^T \hat{\mathbf{e}} = (\mathbf{Y} - \hat{\mathbf{Y}})^T (\mathbf{Y} - \hat{\mathbf{Y}})$.

In the context of this book, $\mathbf{X}$ is the $n \times p'$ matrix giving the values of the independent variables. The ith row of $\mathbf{X}$ is denoted by $\mathbf{x}_i^T$. In models with an intercept $\mathbf{x}_i$ is $(p + 1) \times 1$, with a 1 in the first place. When needed, the columns of $\mathbf{X}$ will be denoted by subscripted boldface capital letters so that, for example, $\mathbf{X}_3$ will be the column of $\mathbf{X}$ corresponding to the third independent variable ($\mathbf{X}_0$ will be taken to be a column of ones).

An important matrix obtained from $\mathbf{X}$ is $\mathbf{X}^T \mathbf{X}$. Note that $\mathbf{X}^T \mathbf{X}$ is $p' \times p'$ and the (i, j)-th element is $\mathbf{X}_i^T \mathbf{X}_j = \mathbf{X}_j^T \mathbf{X}_i$, which is the (j, i)-th element of $\mathbf{X}^T \mathbf{X}$. Hence, $\mathbf{X}^T \mathbf{X}$ is symmetric.

Partitioned matrix. Occasionally, it is convenient to have a notation for a part of a matrix. For example, an $n \times p'$ matrix $\mathbf{X}$ can be partitioned by columns into two matrices $\mathbf{X}_1$ and $\mathbf{X}_2$ where

$$\mathbf{X} = (\mathbf{X}_1 \; \mathbf{X}_2) \tag{2A.2}$$

and $\mathbf{X}_1$ is the first q columns of $\mathbf{X}$ and $\mathbf{X}_2$ is the last $p' - q$ columns of $\mathbf{X}$. The matrix product $\mathbf{X}^T \mathbf{X}$ is

$$\mathbf{X}^T \mathbf{X} = (\mathbf{X}_1 \; \mathbf{X}_2)^T (\mathbf{X}_1 \; \mathbf{X}_2)$$

$$= \begin{pmatrix} \mathbf{X}_1^T \mathbf{X}_1 & \mathbf{X}_1^T \mathbf{X}_2 \\ \mathbf{X}_2^T \mathbf{X}_1 & \mathbf{X}_2^T \mathbf{X}_2 \end{pmatrix}$$

Inverse of a matrix. Suppose we have a $p \times p$ matrix $\mathbf{C}$. If we can find another $p \times p$ matrix, say $\mathbf{D}$, such that $\mathbf{CD} = \mathbf{DC} = \mathbf{I}_p$, then we shall say that C is invertable (has an inverse), and that the inverse of C, written C^{-1}, is equal to D. If the inverse exists, it is unique.

Computationally, finding an inverse can be a very difficult matter, except in special circumstances. For example, the inverse of the identity matrix I_p is itself, I_p, as can be verified by direct multiplication ($I_p I_p = I_p$). Also, suppose that C is a $p \times p$ diagonal matrix, say

$$C = \begin{pmatrix} 3 & 0 & 0 & 0 \\ 0 & -1 & 0 & 0 \\ 0 & 0 & 4 & 0 \\ 0 & 0 & 0 & 1 \end{pmatrix}$$

Then the inverse of C is simply that matrix C^{-1} that is a diagonal matrix with diagonal entries that are the reciprocals of the diagonal entries of C, namely,

$$C^{-1} = \begin{pmatrix} \frac{1}{3} & 0 & 0 & 0 \\ 0 & -1 & 0 & 0 \\ 0 & 0 & \frac{1}{4} & 0 \\ 0 & 0 & 0 & 1 \end{pmatrix}$$

It is easily verified that $C^{-1}C = CC^{-1} = I_4$. This method will work for inverting any diagonal matrix, as long as none of the diagonal elements is zero. If there are zeros on the diagonal, then no inverse exists.

The most important class of matrices for which finding an inverse is easy are the orthogonal matrices. An $n \times n$ matrix Q is *orthogonal* if $Q^T Q = QQ^T = I_n$. Thus $Q^{-1} = Q^T$. For example, the matrix

$$Q = \begin{pmatrix} \dfrac{1}{\sqrt{3}} & \dfrac{1}{\sqrt{2}} & \dfrac{1}{\sqrt{6}} \\ \dfrac{1}{\sqrt{3}} & 0 & -\dfrac{2}{\sqrt{6}} \\ \dfrac{1}{\sqrt{3}} & -\dfrac{1}{\sqrt{2}} & \dfrac{1}{\sqrt{6}} \end{pmatrix}$$

is orthogonal,

$$Q^T = Q^{-1} = \begin{pmatrix} \dfrac{1}{\sqrt{3}} & \dfrac{1}{\sqrt{3}} & \dfrac{1}{\sqrt{3}} \\ \dfrac{1}{\sqrt{2}} & 0 & -\dfrac{1}{\sqrt{2}} \\ \dfrac{1}{\sqrt{6}} & -\dfrac{2}{\sqrt{6}} & \dfrac{1}{\sqrt{6}} \end{pmatrix}$$

since $QQ^T = I$.

In most regression problems where matrix inversion is required, the best approach is to change the original problem into an equivalent one for which finding the inverse is either easy or not actually necessary. An example of this approach is outlined in Appendix 2A.3.

Rank of a matrix. Not all square matrices will have inverses. This is analogous to the observation that not all real numbers have inverses (for real numbers only zero has no inverse; the inverse of any real number k is $k^{-1} = 1/k$, so that, for example, if $k = 3, k^{-1} = \frac{1}{3}$, since $kk^{-1} = 3(\frac{1}{3}) = 1$). If a square matrix has an inverse, we shall say that it is of *full rank* or *invertable*, or *nonsingular*, three terms that are used interchangeably in this book.

To illustrate matrix manipulations, we will substitute $\hat{\beta} = (\mathbf{X}^T\mathbf{X})^{-1}\mathbf{X}^T\mathbf{Y}$ into $RSS = (\mathbf{Y} - \mathbf{X}\hat{\beta})^T(\mathbf{Y} - \mathbf{X}\hat{\beta})$ and simplify. First, performing the indicated multiplications, RSS is

$$RSS = (\mathbf{Y} - \mathbf{X}\hat{\beta})^T(\mathbf{Y} - \mathbf{X}\hat{\beta})$$
$$= \mathbf{Y}^T\mathbf{Y} - \hat{\beta}^T\mathbf{X}^T\mathbf{Y} - \mathbf{Y}^T\mathbf{X}\hat{\beta} + \hat{\beta}^T\mathbf{X}^T\mathbf{X}\hat{\beta} \tag{2A.3}$$

All terms on the right hand side of (2A.3) are 1×1, so $(\mathbf{Y}^T\mathbf{X}\hat{\beta}) = (\mathbf{Y}^T\mathbf{X}\hat{\beta})^T = \hat{\beta}^T\mathbf{X}^T\mathbf{Y}$ and

$$RSS = \mathbf{Y}^T\mathbf{Y} - 2\hat{\beta}^T\mathbf{X}^T\mathbf{Y} + \hat{\beta}^T\mathbf{X}^T\mathbf{X}\hat{\beta} \tag{2A.4}$$

Substituting $(\mathbf{X}^T\mathbf{X})^{-1}\mathbf{X}^T\mathbf{Y}$ for $\hat{\beta}$ in the last term of (2A.4),

$$RSS = \mathbf{Y}^T\mathbf{Y} - 2\hat{\beta}\mathbf{X}^T\mathbf{Y} + \hat{\beta}^T(\mathbf{X}^T\mathbf{X})(\mathbf{X}^T\mathbf{X})^{-1}\mathbf{X}^T\mathbf{Y}$$
$$= \mathbf{Y}^T\mathbf{Y} - 2\hat{\beta}\mathbf{X}^T\mathbf{Y} + \hat{\beta}\mathbf{I}_p\mathbf{X}^T\mathbf{Y}$$
$$= \mathbf{Y}^T\mathbf{Y} - \hat{\beta}^T\mathbf{X}^T\mathbf{Y} \tag{2A.5}$$

since $\hat{\beta}^T\mathbf{I}_p\mathbf{X}^T\mathbf{Y} = \hat{\beta}^T\mathbf{X}^T\mathbf{Y}$. We can continue by substituting for $\hat{\beta}^T$ in (2A.5) to get

$$RSS = \mathbf{Y}^T\mathbf{Y} - \left[(\mathbf{X}^T\mathbf{X})^{-1}\mathbf{X}^T\mathbf{Y}\right]^T\mathbf{X}^T\mathbf{Y}$$
$$= \mathbf{Y}^T\mathbf{Y} - \mathbf{Y}^T\mathbf{X}(\mathbf{X}^T\mathbf{X})^{-1}\mathbf{X}^T\mathbf{Y}$$

This last result requires recognizing that $(\mathbf{X}^T\mathbf{X})^{-T}$ (e.g., the inverse of the transpose) equals $(\mathbf{X}^T\mathbf{X})^{-1}$ because $\mathbf{X}^T\mathbf{X}$ is symmetric. Similar manipulations will give the other forms of (2.22).

If a computer were being programmed to compute RSS, one would compute $\mathbf{Y}^T\mathbf{Y}$, $\hat{\beta}$, and $\mathbf{X}^T\mathbf{Y}$ (or the analogous forms corrected for the sample average). Then, compute $\hat{\beta}^T(\mathbf{X}^T\mathbf{Y})$, and subtract from $\mathbf{Y}^T\mathbf{Y}$ to get RSS.

2A.2   Random vectors

A vector whose elements are random variables is called a *random vector*. In regression, the $n \times 1$ vector $\mathbf{e}$ of errors is a random vector. Other important random vectors in regression are the estimated parameter vector $\hat{\beta}$, the vector of observed values $\mathbf{Y}$ and of fitted values $\hat{\mathbf{Y}}$, and the residual vector $\hat{\mathbf{e}}$.

The mean or expected value of a random vector is the vector of means of the random variables in that vector. Thus, for example, the mean of $\mathbf{e}$ is a vector of all zeros, assuming the $E(e_i) = 0$. We write this as $E(\mathbf{e}) = \mathbf{0}$.

As in Appendix 1A.2, the mean is linear, which says that if $\mathbf{z}$ is a random vector of size $n \times 1$, and $\mathbf{C}$ is any $q \times n$ matrix, and $\mathbf{d}$ any $q \times 1$ fixed vector, then the mean of the random variable $\mathbf{Cz} + \mathbf{d}$ is $E(\mathbf{Cz} + \mathbf{d}) = \mathbf{C}E(\mathbf{z}) + \mathbf{d}$. We can apply this rule to find the mean of $\hat{\beta}$,

$$E(\hat{\beta}) = E\left[(\mathbf{X}^T\mathbf{X})^{-1}\mathbf{X}^T\mathbf{Y}\right] = E\left[(\mathbf{X}^T\mathbf{X})^{-1}\mathbf{X}^T(\mathbf{X}\beta + \mathbf{e})\right]$$

since, according to the model, $\mathbf{Y} = \mathbf{X}\beta + \mathbf{e}$. Then,

$$E(\hat{\beta}) = (\mathbf{X}^T\mathbf{X})^{-1}\mathbf{X}^T\mathbf{X}\beta + (\mathbf{X}^T\mathbf{X})^{-1}\mathbf{X}^T E(\mathbf{e}) \qquad (2A.6)$$

But, since $E(\mathbf{e}) = \mathbf{0}$,

$$E(\hat{\beta}) = (\mathbf{X}^T\mathbf{X})^{-1}\mathbf{X}^T\mathbf{X}\beta = \beta$$

and $\hat{\beta}$ is an unbiased estimate of β. Incidentally, this shows exactly what the bias in the model would be if the model were not correct and if $E(\mathbf{e}) \neq \mathbf{0}$. Suppose, for example, that $E(\mathbf{e}) = \mathbf{Z}\gamma$ for some $n \times q$ matrix $\mathbf{Z}$ of variables different from those in $\mathbf{X}$ (although $\mathbf{Z}$ may be transformations or combinations of the columns of $\mathbf{X}$), and γ is a $q \times 1$ (unknown) parameter vector. Then

$$E(\hat{\beta}) = \beta + (\mathbf{X}^T\mathbf{X})^{-1}\mathbf{X}^T\mathbf{Z}\gamma$$

and the bias in $\hat{\beta}$ as an estimate of β is given by

$$\text{bias} = \beta - E(\hat{\beta}) = -(\mathbf{X}^T\mathbf{X})^{-1}\mathbf{X}^T\mathbf{Z}\gamma \qquad (2A.7)$$

The matrix $(\mathbf{X}^T\mathbf{X})^{-1}\mathbf{X}^T\mathbf{Z}$ is called the *alias* matrix, as each element in $\hat{\beta}$ is confused or aliased with the elements in γ in a way that is determined by the alias matrix. The bias will be zero only if (1) the product $\mathbf{X}^T\mathbf{Z} = \mathbf{0}$ (the columns of $\mathbf{Z}$ are then orthogonal to the columns of $\mathbf{X}$) or (2) if $\gamma = \mathbf{0}$. This is the exact situation that will arise if we fit the model $\mathbf{Y} = \mathbf{X}\beta + \mathbf{e}$, but the true model is actually $\mathbf{Y} = \mathbf{X}\beta + \mathbf{Z}\gamma + \mathbf{e}$. It is left as an exercise to show that, if the smaller model is true, and the larger one is fit, then the estimator of β is unbiased.

Variance-covariance. A random vector has a variance-covariance matrix associated with it. The diagonal entries of this matrix are the variances of the elements of the random vector, while the off-diagonal entries are the covariances between the elements; the (i, j)-th element of the variance covariance matrix is the covariance between the ith element of the random vector and the jth element of the random vector. We use the symbol var($\mathbf{z}$) to denote the variance covariance matrix of the random vector $\mathbf{z}$.

The error vector $\mathbf{e}$ has been assumed to have elements with common variance, and zero covariances. This is summarized as var($\mathbf{e}$) $= \sigma^2 \mathbf{I}_n$. A random vector with unequal variances but uncorrelated elements is given by a diagonal matrix,

$$\begin{bmatrix} \sigma_1^2 & & & \\ & \sigma_2^2 & & 0 \\ 0 & & \ddots & \\ & & & \sigma_n^2 \end{bmatrix}$$

The formula for var($\mathbf{Cz} + \mathbf{d}$) is given by

$$\text{var}(\mathbf{Cz} + \mathbf{d}) = \mathbf{C}\big[\text{var}(\mathbf{z})\big]\mathbf{C}^T$$

Applying this to

$$\hat{\boldsymbol{\beta}} = (\mathbf{X}^T\mathbf{X})^{-1}\mathbf{X}^T\mathbf{Y} = (\mathbf{X}^T\mathbf{X})^{-1}\mathbf{X}^T(\mathbf{X}\boldsymbol{\beta} + \mathbf{e})$$

$$= (\mathbf{X}^T\mathbf{X})^{-1}\mathbf{X}^T\mathbf{X}\boldsymbol{\beta} + (\mathbf{X}^T\mathbf{X})^{-1}\mathbf{X}^T\mathbf{e}$$

note that the first term does not involve $\mathbf{e}$, and therefore corresponds to $\mathbf{d}$ in the formula. Thus, associating $\mathbf{C}$ with $(\mathbf{X}^T\mathbf{X})^{-1}\mathbf{X}^T$,

$$\text{var}(\hat{\boldsymbol{\beta}}) = (\mathbf{X}^T\mathbf{X})^{-1}\mathbf{X}^T\big[\text{var}(\mathbf{e})\big]\big[(\mathbf{X}^T\mathbf{X})^{-1}\mathbf{X}^T\big]^T$$

$$= (\mathbf{X}^T\mathbf{X})^{-1}\mathbf{X}^T(\sigma^2\mathbf{I})\mathbf{X}(\mathbf{X}^T\mathbf{X})^{-1}$$

$$= \sigma^2(\mathbf{X}^T\mathbf{X})^{-1}(\mathbf{X}^T\mathbf{X})(\mathbf{X}^T\mathbf{X})^{-1}$$

$$= \sigma^2(\mathbf{X}^T\mathbf{X})^{-1}$$

as given by (2.20).

Another important application of this result is to find the variance of the fitted value corresponding to the $(p + 1) \times 1$ vector $\mathbf{x}$. The fitted value is given by $\hat{y} = \mathbf{x}^T\hat{\boldsymbol{\beta}}$ and therefore

$$\text{var}(\hat{y}) = \mathbf{x}^T\big[\text{var}(\hat{\boldsymbol{\beta}})\big]\mathbf{x} = \sigma^2\mathbf{x}^T(\mathbf{X}^T\mathbf{X})^{-1}\mathbf{x} \qquad (2A.8)$$

For a prediction at $\mathbf{x}$, the variance of $\tilde{y} = \mathbf{x}^T\hat{\boldsymbol{\beta}} + e$ is

$$\text{var}(\tilde{y}) = \sigma^2\left[1 + \mathbf{x}^T(\mathbf{X}^T\mathbf{X})^{-1}\mathbf{x}\right] \tag{2A.9}$$

2A.3 Least squares

Rather than repeat the derivation of the least squares estimators from Chapter 1, we shall approach the problem in a different way that not only will lead to finding the least squares estimators, but will also outline an important computational method. Recall that the least squares estimator $\hat{\boldsymbol{\beta}}$ minimizes

$$RSS = (\mathbf{Y} - \mathbf{X}\hat{\boldsymbol{\beta}})^T(\mathbf{Y} - \mathbf{X}\hat{\boldsymbol{\beta}}) \tag{2A.10}$$

Suppose that we could find an $n \times p'$ matrix $\mathbf{Q}_1$ with orthogonal columns (e.g., $\mathbf{Q}_1^T\mathbf{Q}_1 = \mathbf{I}_{p'}$) and a $p' \times p'$ upper triangular matrix $\mathbf{R}$ (i.e., a matrix with all zeros below the main diagonal), such that

$$\mathbf{X} = \mathbf{Q}_1\mathbf{R} \tag{2A.11}$$

Then, given $\mathbf{Q}_1$ and $\mathbf{R}$, the model $\mathbf{Y} = \mathbf{X}\boldsymbol{\beta} + \mathbf{e}$ can be rewritten

$$\mathbf{Y} = \mathbf{X}\boldsymbol{\beta} + \mathbf{e} = \mathbf{Q}_1\mathbf{R}\boldsymbol{\beta} + \mathbf{e} = \mathbf{Q}_1\boldsymbol{\gamma} + \mathbf{e}$$

where now $\boldsymbol{\gamma} = \mathbf{R}\boldsymbol{\beta}$ is the parameter vector in the transformed problem. The least squares problem we shall solve is to find $\hat{\boldsymbol{\gamma}}$ to minimize

$$RSS = (\mathbf{Y} - \mathbf{Q}_1\hat{\boldsymbol{\gamma}})^T(\mathbf{Y} - \mathbf{Q}_1\hat{\boldsymbol{\gamma}}) \tag{2A.12}$$

To solve this problem, suppose that $\mathbf{Q}$ is an $n \times n$ orthogonal matrix with $\mathbf{Q}_1$ as the first p' columns, that is

$$\mathbf{Q} = (\mathbf{Q}_1 \, \mathbf{Q}_2) \tag{2A.13}$$

Since $\mathbf{Q}$ is orthogonal,

$$\mathbf{I}_n = \mathbf{Q}\mathbf{Q}^T = (\mathbf{Q}_1 \, \mathbf{Q}_2)\begin{pmatrix} \mathbf{Q}_1^T \\ \mathbf{Q}_2^T \end{pmatrix} = \mathbf{Q}_1^T\mathbf{Q}_1 + \mathbf{Q}_2^T\mathbf{Q}_2$$

and

$$\mathbf{I}_n = \mathbf{Q}^T\mathbf{Q} = \begin{pmatrix} \mathbf{Q}_1^T \\ \mathbf{Q}_2^T \end{pmatrix}(\mathbf{Q}_1 \, \mathbf{Q}_2) = \begin{pmatrix} \mathbf{Q}_1^T\mathbf{Q}_1 & \mathbf{Q}_1^T\mathbf{Q}_2 \\ \mathbf{Q}_2^T\mathbf{Q}_1 & \mathbf{Q}_2^T\mathbf{Q}_2 \end{pmatrix} \tag{2A.14}$$

so in particular $\mathbf{Q}_1^T\mathbf{Q}_1 = \mathbf{I}_{p'}$, $\mathbf{Q}_2^T\mathbf{Q}_2 = \mathbf{I}_{n-p'}$, $\mathbf{Q}_1^T\mathbf{Q}_2 = \mathbf{0}$, and $\mathbf{Q}_2^T\mathbf{Q}_1 = \mathbf{0}$. Now, we can write

$$\mathbf{Y} = \mathbf{I}_n\mathbf{Y} = \mathbf{Q}\mathbf{Q}^T\mathbf{Y} = \mathbf{Q}_1\mathbf{Q}_1^T\mathbf{Y} + \mathbf{Q}_2\mathbf{Q}_2^T\mathbf{Y} \tag{2A.15}$$

Then, substituting (2A.15) into (2A.12),

$$RSS = (Q_1Q_1^TY + Q_2Q_2^TY - Q_1\hat{\gamma})^T(Q_1Q_1^TY + Q_2Q_2^TY - Q_1\hat{\gamma})$$

$$= [Q_2Q_2^TY - Q_1(Q_1^TY - \hat{\gamma})]^T[Q_2Q_2^TY - Q_1(Q_1^TY - \hat{\gamma})]$$

Multiplying this out and using (2A.14), two of the four resulting terms are zero, and the remaining quantity is

$$RSS = Y^TQ_2Q_2^TY + (Q_1^TY - \hat{\gamma})^T(Q_1^TY - \hat{\gamma}) \qquad (2A.16)$$

The least squares estimator is now immediate, since the first term is independent of $\hat{\gamma}$, and the second is made zero by setting

$$\hat{\gamma} = Q_1^TY \qquad (2A.17)$$

As a bonus, we see immediately that $RSS = Y^TQ_2Q_2^TY$.

The least squares estimator $\hat{\beta}$ is obtained by solving

$$R\hat{\beta} = \hat{\gamma} \qquad (2A.18)$$

so that, assuming R^{-1} exists,

$$\hat{\beta} = R^{-1}\hat{\gamma} = R^{-1}Q_1^TY \qquad (2A.19)$$

That (2A.19) is equivalent to (2.17) is left as an exercise.

In practice, it is unnecessary to actually invert R to obtain $\hat{\beta}$. Rather, it may obtained by *back substitution*. Suppose, for example, that R and $z_1 = Q_1^TY$ were given by

$$R = \begin{bmatrix} 1 & 2 & 3 \\ 0 & 4 & 5 \\ 0 & 0 & 6 \end{bmatrix} \qquad z_1 = \begin{bmatrix} 2 \\ 1 \\ 3 \end{bmatrix}$$

We want to solve

$$\begin{bmatrix} 1 & 2 & 3 \\ 0 & 4 & 5 \\ 0 & 0 & 6 \end{bmatrix} \begin{bmatrix} \hat{\beta}_0 \\ \hat{\beta}_1 \\ \hat{\beta}_2 \end{bmatrix} = \begin{bmatrix} 2 \\ 1 \\ 3 \end{bmatrix}$$

for $\hat{\beta}_0$, $\hat{\beta}_1$, and $\hat{\beta}_2$—a system of three equations in three unknowns. These equations are

$$1\hat{\beta}_0 + 2\hat{\beta}_1 + 3\hat{\beta}_2 = 2$$

$$0 + 4\hat{\beta}_1 + 5\hat{\beta}_2 = 1$$

$$0 + 0 + 6\hat{\beta}_2 = 3$$

Solving the bottom equation first gives $\hat{\beta}_2 = \frac{3}{6} = \frac{1}{2}$. Substituting in the middle equation, $4\hat{\beta}_1 + 5(\frac{1}{2}) = 1$ or $\hat{\beta}_1 = -\frac{3}{8}$, and $\hat{\beta}_0 + 2(-\frac{3}{8}) + 3(\frac{1}{2}) = 1$, or $\hat{\beta}_0 = \frac{1}{4}$. This method will work for any triangular system of equations.

The QR factorization used here provides a numerically stable algorithm for doing least squares computations. A method for actually finding $\mathbf{Q}_1$ and $\mathbf{R}$ is described in books by Chambers (1977), Stewart (1974), Lawson and Hansen (1974), Seber (1977), and Dongarra et al. (1979). Also, FORTRAN subroutines for the QR decomposition are available in the LINPACK package documented in the last reference and available for purchase from the International Mathematics and Statistical Library.*

5A.1 Relating regression equations

The case statistics that we examine are practical because of the relationship between regression based on a full sample and regressions based on the sample with one of the cases removed from the data. Actually, the results are more general, and formulas can be developed for adding or deleting any number of cases to a regression problem (although, for some computations, the formulas for deletion are not numerically stable). Expanded presentations of the material given here are available in many sources, including Plackett (1950). The computer manual for the LINPACK collection of computer subroutines (Dongarra et al., 1979) includes well-documented computer codes for relating regression equations.

Suppose that we have a data matrix $\mathbf{X}$ with n rows and p' columns, and a vector $\mathbf{Y}$, and that the matrix $(\mathbf{X}^T\mathbf{X})^{-1}$ has been computed. Now suppose that we want to delete the ith case, and get an $(n-1) \times p'$ matrix $\mathbf{X}_{-i}$ and we want to compute $(\mathbf{X}_{-i}^T\mathbf{X}_{-i})^{-1}$, based on $(\mathbf{X}^T\mathbf{X})^{-1}$. To do this computation, we use the following identity:

$$\left(\mathbf{X}_{-i}^T\mathbf{X}_{-i}\right)^{-1} = \left(\mathbf{X}^T\mathbf{X}\right)^{-1} + \frac{\left(\mathbf{X}^T\mathbf{X}\right)^{-1}\mathbf{x}_i\mathbf{x}_i^T\left(\mathbf{X}^T\mathbf{X}\right)^{-1}}{1 - v_{ii}} \qquad (5\text{A}.1)$$

This remarkable formula can be applied to give all of the results that one would want relating the regressions with and without the ith case. For example, the distance from $\mathbf{x}_i$ to the center of the remaining $n-1$ cases is defined by $\mathbf{x}_i^T(\mathbf{X}_{-i}^T\mathbf{X}_{-i})^{-1}\mathbf{x}_i$. Using (5A.1),

$$\mathbf{x}_i^T\left(\mathbf{X}_{-i}^T\mathbf{X}_{-i}\right)^{-1}\mathbf{x}_i = \mathbf{x}_i^T\left(\mathbf{X}^T\mathbf{X}\right)^{-1}\mathbf{x}_i + \frac{\mathbf{x}_i^T\left(\mathbf{X}^T\mathbf{X}\right)^{-1}\mathbf{x}_i\mathbf{x}_i^T\left(\mathbf{X}^T\mathbf{X}\right)^{-1}\mathbf{x}_i}{1 - v_{ii}}$$

$$= v_{ii} + \frac{v_{ii}^2}{1 - v_{ii}}$$

$$= \frac{v_{ii}}{1 - v_{ii}} \qquad (5\text{A}.2)$$

* Write IMSL, Inc., Sixth Floor, GNB Building, 7500 Bellaire Blvd., Houston, Texas 77036.

which is the result given in the text. Similar computations can be done to find any statistics for the regression deleting the ith case from the regression based on all of the cases.

Estimate of β:

$$\hat{\beta}_{-i} = \hat{\beta} - \frac{(\mathbf{X}^T\mathbf{X})^{-1}\mathbf{x}_i\hat{e}_i}{1 - v_{ii}} \tag{5A.3}$$

Estimate of σ^2:

$$\hat{\sigma}^2_{-i} = \frac{1}{n - p' - 1} \hat{\sigma}^2(n - p' - r_i^2) \tag{5A.4}$$

8A.1 Derivation of C_p

A fixed subset model is specified by a partition of $\mathbf{X} = (\mathbf{X}_1\ \mathbf{X}_2)$ so that the subset model is

$$\mathbf{Y} = \mathbf{X}_1\boldsymbol{\beta}_1 + \mathbf{e} \qquad \text{var}(\mathbf{e}) = \sigma^2\mathbf{I}$$

Now, suppose that for the subset model the ith fitted value is $\hat{y}_i$, and let $\mathbf{V} = \mathbf{X}_1(\mathbf{X}_1^T\mathbf{X}_1)^{-1}\mathbf{X}_1^T$, so, if v_{ii} is the ith diagonal element of $\mathbf{V}$, then $\text{var}(\hat{y}_i) = \sigma^2 v_{ii}$. Similarly, for the full model, let $\hat{Y}_i$ be the ith fitted value, $\mathbf{U} = \mathbf{X}(\mathbf{X}^T\mathbf{X})^{-1}\mathbf{X}^T$, with diagonal entries u_{ii} and $\text{var}(\hat{Y}_i) = \sigma^2 u_{ii}$.

Now, as in text, define

$$E(J_p) = \frac{1}{\sigma^2} \sum_{i=1}^{n} \text{mse}(\hat{y}_i)$$

where

$$\text{mse}(\hat{y}_i) = \sigma^2 v_{ii} + \left[E(\hat{y}_i) - E(y_i) \right]^2$$

If the full model (8.7) is unbiased, $E(\hat{Y}_i) = E(y_i)$, one can show that

$$E\left[\hat{y}_i - E(y_i) \right]^2 = \left[E(\hat{y}_i) - E(\hat{Y}_i) \right]^2 + \sigma^2(u_{ii} - v_{ii})$$

(This last result is most easily shown if $\mathbf{X}$ is first transformed via a QR factorization, as left to the interested reader.) Thus we can write

$$\text{mse}(\hat{y}_i) = E(\hat{y}_i - \hat{Y}_i)^2 + \sigma^2 v_{ii} - \sigma^2(u_{ii} - v_{ii})$$

Now, again assuming the full model to be unbiased, $\hat{\sigma}^2$ from the full model estimates σ^2, and $(\hat{y}_i - \hat{Y}_i)^2$ estimates $E(\hat{y}_i - \hat{Y}_i)^2$. Substituting, we get an estimate $\widehat{\text{mse}}(\hat{y}_i)$,

$$\widehat{\text{mse}}(\hat{y}_i) = (\hat{y}_i - \hat{Y}_i)^2 + \hat{\sigma}^2\left[v_{ii} - (u_{ii} - v_{ii}) \right] \tag{8A.1}$$

So an estimate of J_p is C_p, where

$$C_p = \frac{1}{\hat{\sigma}^2} \sum_{i=1}^{n} \widehat{\text{mse}}(\hat{y}_i)$$

$$= \sum_{i=1}^{n} \frac{(\hat{y}_i - \hat{Y}_i)^2}{\hat{\sigma}^2} + v_{ii} - (u_{ii} - v_{ii}) \qquad (8A.2)$$

The terms $\widehat{\text{mse}}(\hat{y}_i)/\hat{\sigma}^2$ can serve as a case statistic for assessing subset models, as it gives the estimated MSE/σ^2 for each case separately (Weisberg, 1979).

To get the usual form for C_p, we need three results, again left for proof to the interested reader.

1. $\sum(\hat{y}_i - \hat{Y}_i)^2 =$ additional sum of squares for regression on X_2 after X_1.

2. $\sum v_{ii} = p$.

3. $\sum u_{ii} = k'$.

Substituting these into (8A.2) gives (8.15).

9A.1   Finding a minimum content ellipse

Since computer programs for finding the MCE for prediction problems are not readily available, one algorithm for performing the necessary calculations is presented here. This is an iterative algorithm based on the theory of optimal experimental designs, and is adapted from Federov (1972); other algorithms are discussed by Titterington (1978).

Suppose we let $\mathbf{x}_1, \ldots, \mathbf{x}_n$ be the n predictor vectors in the construction sample, where each $\mathbf{x}_i$ is $p \times 1$. Although the model we use has an intercept term, the constant 1 does not appear in each $\mathbf{x}_i$. At the kth step of the algorithm, suppose we have computed a $p \times 1$ vector $\mathbf{m}_k$ and a $p \times p$ matrix $\mathbf{M}_k$. Then, we compute the quantity

$$c_k = \max_{i} (\mathbf{x}_i - \mathbf{m}_k)^T \mathbf{M}_k^{-1} (\mathbf{x}_i - \mathbf{m}_k)$$

This requires a separate computation for all n data points. We will then stop the iteration if $c_k < p + \epsilon$, where ϵ is a small number; $\epsilon = 0.1$ appears as a good choice. If $c_k \geqslant p + \epsilon$ we must iterate more by updating $\mathbf{m}_k$ and $\mathbf{M}_k$ to $\mathbf{m}_{k+1}$ and $\mathbf{M}_{k+1}$. Suppose that c_k occurs at the point $\mathbf{x}_l$ (if c_k occurs at several points, use any one of them). Then we let $\alpha = (c_k - p)/p(c_k - 1)$

and update $\mathbf{m}_k$ and $\mathbf{M}_k$ as

$$\mathbf{m}_{k+1} = (1 - \alpha)\mathbf{m}_k + \alpha\mathbf{x}_l$$

$$\mathbf{M}_{k+1} = (1 - \alpha)\mathbf{M}_k + \alpha\mathbf{x}_l\mathbf{x}_l^T$$

Then, compute c_{k+1} and repeat the iteration until convergence.

To begin the process, it is convenient to take $\mathbf{m}_0 = \sum\mathbf{x}_i/n = $ sample average vector, $\mathbf{M}_0 = (\mathcal{X}^T\mathcal{X})/n = $ corrected cross product matrix divided by n. Then the first ellipsoid will be the same as that given by the contours of the v_{ii} as discussed in the text.

TABLES

Table A   Student's *t*-distribution

The tabled values are two-tailed values $t(\alpha; \nu)$, such that

$$\text{prob}\{|t_\nu \text{ variate}| > t(\alpha; \nu)\} = \alpha$$

The entries in the table were computed on a CDC Cyber 172 computer at the University of Minnesota using IMSL subroutine MDSTI

	α				
ν	0.200	0.100	0.050	0.010	0.001
1	3.08	6.31	12.71	63.66	636.62
2	1.89	2.92	4.30	9.92	31.60
3	1.64	2.35	3.18	5.84	12.92
4	1.53	2.13	2.78	4.60	8.61
5	1.48	2.02	2.57	4.03	6.87
6	1.44	1.94	2.45	3.71	5.96
7	1.41	1.89	2.36	3.50	5.41
8	1.40	1.86	2.31	3.36	5.04
9	1.38	1.83	2.26	3.25	4.78
10	1.37	1.81	2.23	3.17	4.59
11	1.36	1.80	2.20	3.11	4.44
12	1.36	1.78	2.18	3.05	4.32
13	1.35	1.77	2.16	3.01	4.22
14	1.35	1.76	2.14	2.98	4.14
15	1.34	1.75	2.13	2.95	4.07
16	1.34	1.75	2.12	2.92	4.01
17	1.33	1.74	2.11	2.90	3.97
18	1.33	1.73	2.10	2.88	3.92
19	1.33	1.73	2.09	2.86	3.88
20	1.33	1.72	2.09	2.85	3.85
21	1.32	1.72	2.08	2.83	3.82
22	1.32	1.72	2.07	2.82	3.79
23	1.32	1.71	2.07	2.81	3.77

ν	α				
	0.200	0.100	0.050	0.010	0.001
24	1.32	1.71	2.06	2.80	3.75
25	1.32	1.71	2.06	2.79	3.73
26	1.31	1.71	2.06	2.78	3.71
27	1.31	1.70	2.05	2.77	3.69
28	1.31	1.70	2.05	2.76	3.67
29	1.31	1.70	2.05	2.76	3.66
30	1.31	1.70	2.04	2.75	3.65
31	1.31	1.70	2.04	2.74	3.63
32	1.31	1.69	2.04	2.74	3.62
33	1.31	1.69	2.03	2.73	3.61
34	1.31	1.69	2.03	2.73	3.60
35	1.31	1.69	2.03	2.72	3.59
36	1.31	1.69	2.03	2.72	3.58
37	1.30	1.69	2.03	2.72	3.57
38	1.30	1.69	2.02	2.71	3.57
39	1.30	1.68	2.02	2.71	3.56
40	1.30	1.68	2.02	2.70	3.55
41	1.30	1.68	2.02	2.70	3.54
42	1.30	1.68	2.02	2.70	3.54
43	1.30	1.68	2.02	2.70	3.53
44	1.30	1.68	2.02	2.69	3.53
45	1.30	1.68	2.01	2.69	3.52
46	1.30	1.68	2.01	2.69	3.51
47	1.30	1.68	2.01	2.68	3.51
48	1.30	1.68	2.01	2.68	3.51
49	1.30	1.68	2.01	2.68	3.50
50	1.30	1.68	2.01	2.68	3.50
60	1.30	1.67	2.00	2.66	3.46
70	1.29	1.67	1.99	2.65	3.44
80	1.29	1.66	1.99	2.64	3.42
90	1.29	1.66	1.99	2.63	3.40
100	1.29	1.66	1.98	2.63	3.39
120	1.29	1.66	1.98	2.62	3.37
∞	1.28	1.64	1.96	2.58	3.29

Table B F-distribution

The tabled values are $F(\alpha, \nu_1, \nu_2)$ such that $\text{prob}\{F(\nu_1, \nu_2) \text{ variate} > F(\alpha, \nu_1, \nu_2)\} = \alpha$.

$$\alpha = 0.05$$

Degrees of freedom for numerator

ν_2 \ ν_1	1	2	3	4	5	6	7	8	9	10	12	15	20	24	30	40	60	120	∞
1	161.4	199.5	215.7	224.6	230.2	234.0	236.8	238.9	240.5	241.9	243.9	245.9	248.0	249.1	250.1	251.1	252.2	253.3	254.3
2	18.51	19.00	19.16	19.25	19.30	19.33	19.35	19.37	19.38	19.40	19.41	19.43	19.45	19.45	19.46	19.47	19.48	19.49	19.50
3	10.13	9.55	9.28	9.12	9.01	8.94	8.89	8.85	8.81	8.79	8.74	8.70	8.66	8.64	8.62	8.59	8.57	8.55	8.53
4	7.71	6.94	6.59	6.39	6.26	6.16	6.09	6.04	6.00	5.96	5.91	5.86	5.80	5.77	5.75	5.72	5.69	5.66	5.63
5	6.61	5.79	5.41	5.19	5.05	4.95	4.88	4.82	4.77	4.74	4.68	4.62	4.56	4.53	4.50	4.46	4.43	4.40	4.36
6	5.99	5.14	4.76	4.53	4.39	4.28	4.21	4.15	4.10	4.06	4.00	3.94	3.87	3.84	3.81	3.77	3.74	3.70	3.67
7	5.59	4.74	4.35	4.12	3.97	3.87	3.79	3.73	3.68	3.64	3.57	3.51	3.44	3.41	3.38	3.34	3.30	3.27	3.23
8	5.32	4.46	4.07	3.84	3.69	3.58	3.50	3.44	3.39	3.35	3.28	3.22	3.15	3.12	3.08	3.04	3.01	2.97	2.93
9	5.12	4.26	3.86	3.63	3.48	3.37	3.29	3.23	3.18	3.14	3.07	3.01	2.94	2.90	2.86	2.83	2.79	2.75	2.71
10	4.96	4.10	3.71	3.48	3.33	3.22	3.14	3.07	3.02	2.98	2.91	2.85	2.77	2.74	2.70	2.66	2.62	2.58	2.54
11	4.84	3.98	3.59	3.36	3.20	3.09	3.01	2.95	2.90	2.85	2.79	2.72	2.65	2.61	2.57	2.53	2.49	2.45	2.40
12	4.75	3.89	3.49	3.26	3.11	3.00	2.91	2.85	2.80	2.75	2.69	2.62	2.54	2.51	2.47	2.43	2.38	2.34	2.30
13	4.67	3.81	3.41	3.18	3.03	2.92	2.83	2.77	2.71	2.67	2.60	2.53	2.46	2.42	2.38	2.34	2.30	2.25	2.21
14	4.60	3.74	3.34	3.11	2.96	2.85	2.76	2.70	2.65	2.60	2.53	2.46	2.39	2.35	2.31	2.27	2.22	2.18	2.13
15	4.54	3.68	3.29	3.06	2.90	2.79	2.71	2.64	2.59	2.54	2.48	2.40	2.33	2.29	2.25	2.20	2.16	2.11	2.07
16	4.49	3.63	3.24	3.01	2.85	2.74	2.66	2.59	2.54	2.49	2.42	2.35	2.28	2.24	2.19	2.15	2.11	2.06	2.01
17	4.45	3.59	3.20	2.96	2.81	2.70	2.61	2.55	2.49	2.45	2.38	2.31	2.23	2.19	2.15	2.10	2.06	2.01	1.96
18	4.41	3.55	3.16	2.93	2.77	2.66	2.58	2.51	2.46	2.41	2.34	2.27	2.19	2.15	2.11	2.06	2.02	1.97	1.92
19	4.38	3.52	3.13	2.90	2.74	2.63	2.54	2.48	2.42	2.38	2.31	2.23	2.16	2.11	2.07	2.03	1.98	1.93	1.88
20	4.35	3.49	3.10	2.87	2.71	2.60	2.51	2.45	2.39	2.35	2.28	2.20	2.12	2.08	2.04	1.99	1.95	1.90	1.84
21	4.32	3.47	3.07	2.84	2.68	2.57	2.49	2.42	2.37	2.32	2.25	2.18	2.10	2.05	2.01	1.96	1.92	1.87	1.81
22	4.30	3.44	3.05	2.82	2.66	2.55	2.46	2.40	2.34	2.30	2.23	2.15	2.07	2.03	1.98	1.94	1.89	1.84	1.78
23	4.28	3.42	3.03	2.80	2.64	2.53	2.44	2.37	2.32	2.27	2.20	2.13	2.05	2.01	1.96	1.91	1.86	1.81	1.76
24	4.26	3.40	3.01	2.78	2.62	2.51	2.42	2.36	2.30	2.25	2.18	2.11	2.03	1.98	1.94	1.89	1.84	1.79	1.73
25	4.24	3.39	2.99	2.76	2.60	2.49	2.40	2.34	2.28	2.24	2.16	2.09	2.01	1.96	1.92	1.87	1.82	1.77	1.71
26	4.23	3.37	2.98	2.74	2.59	2.47	2.39	2.32	2.27	2.22	2.15	2.07	1.99	1.95	1.90	1.85	1.80	1.75	1.69
27	4.21	3.35	2.96	2.73	2.57	2.46	2.37	2.31	2.25	2.20	2.13	2.06	1.97	1.93	1.88	1.84	1.79	1.73	1.67
28	4.20	3.34	2.95	2.71	2.56	2.45	2.36	2.29	2.24	2.19	2.12	2.04	1.96	1.91	1.87	1.82	1.77	1.71	1.65
29	4.18	3.33	2.93	2.70	2.55	2.43	2.35	2.28	2.22	2.18	2.10	2.03	1.94	1.90	1.85	1.81	1.75	1.70	1.64
30	4.17	3.32	2.92	2.69	2.53	2.42	2.33	2.27	2.21	2.16	2.09	2.01	1.93	1.89	1.84	1.79	1.74	1.68	1.62
40	4.08	3.23	2.84	2.61	2.45	2.34	2.25	2.18	2.12	2.08	2.00	1.92	1.84	1.79	1.74	1.69	1.64	1.58	1.51
60	4.00	3.15	2.76	2.53	2.37	2.25	2.17	2.10	2.04	1.99	1.92	1.84	1.75	1.70	1.65	1.59	1.53	1.47	1.39
120	3.92	3.07	2.68	2.45	2.29	2.17	2.09	2.02	1.96	1.91	1.83	1.75	1.66	1.61	1.55	1.50	1.43	1.35	1.25
∞	3.84	3.00	2.60	2.37	2.21	2.10	2.01	1.94	1.88	1.83	1.75	1.67	1.57	1.52	1.46	1.39	1.32	1.22	1.00

Degrees of Freedom for Denominator

261

Table B (continued) F-distribution

α = 0.01

Degrees of freedom for numerator

ν_2 \ ν_1	1	2	3	4	5	6	7	8	9	10	12	15	20	24	30	40	60	120	∞
1	4052	4999.5	5403	5625	5764	5859	5928	5982	6022	6056	6106	6157	6209	6235	6261	6287	6313	6339	6366
2	98.50	99.00	99.17	99.25	99.30	99.33	99.36	99.37	99.39	99.40	99.42	99.43	99.45	99.46	99.47	99.47	99.48	99.49	99.50
3	34.12	30.82	29.46	28.71	28.24	27.91	27.67	27.49	27.35	27.23	27.05	26.87	26.69	26.60	26.50	26.41	26.32	26.22	26.13
4	21.20	18.00	16.69	15.98	15.52	15.21	14.98	14.80	14.66	14.55	14.37	14.20	14.02	13.93	13.84	13.75	13.65	13.56	13.46
5	16.26	13.27	12.06	11.39	10.97	10.67	10.46	10.29	10.16	10.05	9.89	9.72	9.55	9.47	9.38	9.29	9.20	9.11	9.02
6	13.75	10.92	9.78	9.15	8.75	8.47	8.26	8.10	7.98	7.87	7.72	7.56	7.40	7.31	7.23	7.14	7.06	6.97	6.88
7	12.25	9.55	8.45	7.85	7.46	7.19	6.99	6.84	6.72	6.62	6.47	6.31	6.16	6.07	5.99	5.91	5.82	5.74	5.65
8	11.26	8.65	7.59	7.01	6.63	6.37	6.18	6.03	5.91	5.81	5.67	5.52	5.36	5.28	5.20	5.12	5.03	4.95	4.86
9	10.56	8.02	6.99	6.42	6.06	5.80	5.61	5.47	5.35	5.26	5.11	4.96	4.81	4.73	4.65	4.57	4.48	4.40	4.31
10	10.04	7.56	6.55	5.99	5.64	5.39	5.20	5.06	4.94	4.85	4.71	4.56	4.41	4.33	4.25	4.17	4.08	4.00	3.91
11	9.65	7.21	6.22	5.67	5.32	5.07	4.89	4.74	4.63	4.54	4.40	4.25	4.10	4.02	3.94	3.86	3.78	3.69	3.60
12	9.33	6.93	5.95	5.41	5.06	4.82	4.64	4.50	4.39	4.30	4.16	4.01	3.86	3.78	3.70	3.62	3.54	3.45	3.36
13	9.07	6.70	5.74	5.21	4.86	4.62	4.44	4.30	4.19	4.10	3.96	3.82	3.66	3.59	3.51	3.43	3.34	3.25	3.17
14	8.86	6.51	5.56	5.04	4.69	4.46	4.28	4.14	4.03	3.94	3.80	3.66	3.51	3.43	3.35	3.27	3.18	3.09	3.00
15	8.68	6.36	5.42	4.89	4.56	4.32	4.14	4.00	3.89	3.80	3.67	3.52	3.37	3.29	3.21	3.13	3.05	2.96	2.87
16	8.53	6.23	5.29	4.77	4.44	4.20	4.03	3.89	3.78	3.69	3.55	3.41	3.26	3.18	3.10	3.02	2.93	2.84	2.75
17	8.40	6.11	5.18	4.67	4.34	4.10	3.93	3.79	3.68	3.59	3.46	3.31	3.16	3.08	3.00	2.92	2.83	2.75	2.65
18	8.29	6.01	5.09	4.58	4.25	4.01	3.84	3.71	3.60	3.51	3.37	3.23	3.08	3.00	2.92	2.84	2.75	2.66	2.57
19	8.18	5.93	5.01	4.50	4.17	3.94	3.77	3.63	3.52	3.43	3.30	3.15	3.00	2.92	2.84	2.76	2.67	2.58	2.49
20	8.10	5.85	4.94	4.43	4.10	3.87	3.70	3.56	3.46	3.37	3.23	3.09	2.94	2.86	2.78	2.69	2.61	2.52	2.42
21	8.02	5.78	4.87	4.37	4.04	3.81	3.64	3.51	3.40	3.31	3.17	3.03	2.88	2.80	2.72	2.64	2.55	2.46	2.36
22	7.95	5.72	4.82	4.31	3.99	3.76	3.59	3.45	3.35	3.26	3.12	2.98	2.83	2.75	2.67	2.58	2.50	2.40	2.31
23	7.88	5.66	4.76	4.26	3.94	3.71	3.54	3.41	3.30	3.21	3.07	2.93	2.78	2.70	2.62	2.54	2.45	2.35	2.26
24	7.82	5.61	4.72	4.22	3.90	3.67	3.50	3.36	3.26	3.17	3.03	2.89	2.74	2.66	2.58	2.49	2.40	2.31	2.21
25	7.77	5.57	4.68	4.18	3.85	3.63	3.46	3.32	3.22	3.13	2.99	2.85	2.70	2.62	2.54	2.45	2.36	2.27	2.17
26	7.72	5.53	4.64	4.14	3.82	3.59	3.42	3.29	3.18	3.09	2.96	2.81	2.66	2.58	2.50	2.42	2.33	2.23	2.13
27	7.68	5.49	4.60	4.11	3.78	3.56	3.39	3.26	3.15	3.06	2.93	2.78	2.63	2.55	2.47	2.38	2.29	2.20	2.10
28	7.64	5.45	4.57	4.07	3.75	3.53	3.36	3.23	3.12	3.03	2.90	2.75	2.60	2.52	2.44	2.35	2.26	2.17	2.06
29	7.60	5.42	4.54	4.04	3.73	3.50	3.33	3.20	3.09	3.00	2.87	2.73	2.57	2.49	2.41	2.33	2.23	2.14	2.03
30	7.56	5.39	4.51	4.02	3.70	3.47	3.30	3.17	3.07	2.98	2.84	2.70	2.55	2.47	2.39	2.30	2.21	2.11	2.01
40	7.31	5.18	4.31	3.83	3.51	3.29	3.12	2.99	2.89	2.80	2.66	2.52	2.37	2.29	2.20	2.11	2.02	1.92	1.80
60	7.08	4.98	4.13	3.65	3.34	3.12	2.95	2.82	2.72	2.63	2.50	2.35	2.20	2.12	2.03	1.94	1.84	1.73	1.60
120	6.85	4.79	3.95	3.48	3.17	2.96	2.79	2.66	2.56	2.47	2.34	2.19	2.03	1.95	1.86	1.76	1.66	1.53	1.38
∞	6.63	4.61	3.78	3.32	3.02	2.80	2.64	2.51	2.41	2.32	2.18	2.04	1.88	1.79	1.70	1.59	1.47	1.32	1.00

Degrees of Freedom for Denominator

Taken from Draper and Smith (1966) and reproduced with permission from E. S. Pearson and H. O. Hartley (1966),

Table C Percentage points of the chi-squared distribution

Tabled values are $\chi^2(\alpha; n)$ such that
$\text{prob}\{\chi^2(n) \text{ variate} \geqslant \chi^2(\alpha; n)\} = \alpha$

d.f.	$\alpha = 0.05$	$\alpha = 0.01$
1	3.84	6.63
2	5.99	9.21
3	7.81	11.34
4	9.49	13.28
5	11.07	15.09
6	12.59	16.81
7	14.07	18.48
8	15.51	20.09
9	16.92	21.67
10	18.31	23.21
11	19.68	24.73
12	21.03	26.22
13	22.36	27.69
14	23.68	29.14
15	25.00	30.58
20	31.41	37.57
25	37.65	44.31
30	43.77	50.89
40	55.76	63.69
50	67.50	76.15
60	79.08	88.38
70	90.53	100.4
80	101.9	112.3
90	113.1	124.1
100	124.3	135.8

Abridged by permission from Table 8 of E. S. Pearson and H. O. Hartley (1966), *Biometrika Tables for Statisticians*, Vol. 1, 3rd ed., London: Cambridge University.

Table D Critical values for the outlier test

The values in the table in row n and column p' are $t(\alpha/n; n - p')$ for the choices of $\alpha = 0.01$ and 0.05. The layout of the table was suggested by Christopher Bingham. The table was computed on a CDC 6400 computer at the University of Minnesota using IMSL subroutine MDSTI.

Critical values for outlier test, $\alpha = .05$

n / p'	1	2	3	4	5	6	7	8	9	10	11	12	13	14	15	20	25	30
6	4.85	6.23	10.89	76.39														
7	4.38	5.07	6.58	11.77	89.12													
8	4.12	4.53	5.26	6.90	12.59	101.9												
9	3.95	4.22	4.66	5.44	7.18	13.36	114.6											
10	3.83	4.03	4.32	4.77	5.60	7.45	14.09	127.3										
11	3.75	3.90	4.10	4.40	4.88	5.75	7.70	14.78	140.1									
12	3.69	3.81	3.96	4.17	4.49	4.98	5.89	7.94	15.44	152.8								
13	3.65	3.74	3.86	4.02	4.24	4.56	5.08	6.02	8.16	16.08	165.5							
14	3.61	3.69	3.79	3.91	4.07	4.30	4.63	5.16	6.14	8.37	16.69	178.2						
15	3.58	3.65	3.73	3.83	3.95	4.12	4.36	4.70	5.25	6.25	8.58	17.28	191.0					
16	3.56	3.62	3.68	3.77	3.87	4.00	4.17	4.41	4.76	5.33	6.36	8.77	17.85	203.7				
17	3.54	3.59	3.65	3.72	3.80	3.90	4.04	4.21	4.46	4.82	5.40	6.47	8.95	18.40	216.4			
18	3.53	3.57	3.62	3.68	3.75	3.83	3.94	4.08	4.26	4.51	4.88	5.47	6.57	9.13	18.93			
19	3.52	3.56	3.60	3.65	3.71	3.78	3.86	3.97	4.11	4.30	4.55	4.93	5.54	6.67	9.30			
20	3.51	3.54	3.58	3.62	3.67	3.73	3.81	3.89	4.00	4.15	4.33	4.59	4.98	5.60	6.76			
21	3.50	3.53	3.57	3.60	3.65	3.70	3.76	3.83	3.92	4.03	4.18	4.37	4.64	5.03	5.67			
22	3.50	3.52	3.55	3.59	3.63	3.67	3.72	3.78	3.86	3.95	4.06	4.21	4.40	4.68	5.08	280.1		
23	3.49	3.52	3.54	3.57	3.61	3.65	3.69	3.75	3.81	3.88	3.98	4.09	4.24	4.44	4.71	21.41		
24	3.49	3.51	3.53	3.56	3.59	3.63	3.67	3.71	3.77	3.83	3.91	4.00	4.12	4.27	4.47	10.07		
25	3.48	3.50	3.53	3.55	3.58	3.61	3.65	3.69	3.73	3.79	3.85	3.93	4.02	4.14	4.30	7.17		
26	3.48	3.50	3.52	3.54	3.57	3.60	3.63	3.66	3.70	3.75	3.81	3.87	3.95	4.05	4.17	5.95		

Table D (continued)

27	3.48	3.50	3.52	3.54	3.56	3.58	3.61	3.65	3.68	3.72	3.77	3.83	3.89	3.97	4.07	5.29	343.8	407.4
28	3.48	3.50	3.51	3.53	3.55	3.58	3.60	3.63	3.66	3.70	3.74	3.79	3.84	3.91	3.99	4.88	23.63	25.66
29	3.48	3.49	3.51	3.53	3.55	3.57	3.59	3.62	3.64	3.68	3.71	3.76	3.81	3.86	3.93	4.61	10.74	11.34
30	3.48	3.49	3.51	3.52	3.54	3.56	3.58	3.60	3.63	3.66	3.69	3.73	3.77	3.82	3.88	4.42	7.53	7.84
31	3.48	3.49	3.50	3.52	3.54	3.55	3.57	3.59	3.62	3.64	3.67	3.71	3.74	3.79	3.84	4.28	6.18	6.39
32	3.48	3.49	3.50	3.52	3.53	3.55	3.57	3.59	3.61	3.63	3.66	3.69	3.72	3.76	3.80	4.17	5.47	5.62
33	3.48	3.49	3.50	3.52	3.53	3.55	3.56	3.58	3.60	3.62	3.64	3.67	3.70	3.74	3.77	4.08	5.03	5.16
34	3.48	3.49	3.50	3.51	3.53	3.54	3.56	3.57	3.59	3.61	3.63	3.66	3.68	3.71	3.75	4.01	4.74	4.84
35	3.48	3.49	3.50	3.51	3.52	3.54	3.55	3.57	3.58	3.60	3.62	3.64	3.67	3.70	3.73	3.96	4.53	4.62
36	3.48	3.49	3.50	3.51	3.52	3.53	3.55	3.56	3.58	3.60	3.61	3.63	3.66	3.68	3.71	3.91	4.37	
37	3.48	3.49	3.50	3.51	3.52	3.53	3.54	3.56	3.57	3.59	3.61	3.62	3.65	3.67	3.69	3.87	4.26	
38	3.48	3.49	3.50	3.51	3.52	3.53	3.54	3.56	3.57	3.58	3.60	3.62	3.64	3.66	3.68	3.84	4.16	
39	3.49	3.49	3.50	3.51	3.52	3.53	3.54	3.55	3.57	3.58	3.59	3.61	3.63	3.65	3.67	3.81	4.09	
40	3.49	3.49	3.50	3.51	3.52	3.53	3.54	3.55	3.56	3.58	3.59	3.60	3.62	3.64	3.66	3.79	4.03	4.09
50	3.51	3.51	3.51	3.52	3.53	3.53	3.55	3.54	3.55	3.56	3.57	3.57	3.58	3.59	3.60	3.66	3.75	3.88
60	3.53	3.53	3.53	3.54	3.54	3.54	3.56	3.55	3.56	3.57	3.57	3.58	3.58	3.59	3.59	3.62	3.67	3.73
70	3.55	3.55	3.55	3.55	3.56	3.56	3.56	3.56	3.57	3.58	3.59	3.59	3.59	3.59	3.60	3.61	3.64	3.67
80	3.57	3.57	3.57	3.57	3.57	3.58	3.58	3.58	3.58	3.60	3.60	3.60	3.60	3.61	3.61	3.61	3.63	3.66
90	3.58	3.59	3.59	3.59	3.61	3.59	3.59	3.60	3.60	3.61	3.61	3.62	3.62	3.62	3.62	3.62	3.63	3.65
100	3.60	3.60	3.60	3.60	3.61	3.61	3.61	3.61	3.61	3.62	3.62	3.62	3.62	3.63	3.62	3.63	3.64	3.65
200	3.73	3.73	3.73	3.73	3.73	3.73	3.73	3.73	3.73	3.73	3.73	3.73	3.73	3.74	3.74	3.74	3.74	3.74
300	3.81	3.81	3.81	3.81	3.81	3.81	3.81	3.81	3.81	3.81	3.82	3.82	3.82	3.82	3.82	3.82	3.82	3.82
400	3.87	3.87	3.87	3.87	3.87	3.87	3.87	3.87	3.88	3.88	3.88	3.88	3.88	3.88	3.88	3.88	3.88	3.88
500	3.92	3.92	3.92	3.92	3.92	3.92	3.92	3.92	3.92	3.92	3.92	3.92	3.92	3.92	3.92	3.92	3.92	3.92

Table D (continued)

Critical values for outlier test, $\alpha = .01$

n / p'	1	2	3	4	5	6	7	8	9	10	11	12	13	14	15	20	25	30
6	7.53	10.87	24.46	382.0														
7	6.35	7.84	11.45	26.43	445.6													
8	5.71	6.54	8.12	11.98	28.26	509.3												
9	5.31	5.84	6.71	8.38	12.47	29.97	573.0											
10	5.04	5.41	5.96	6.87	8.61	12.92	31.60	636.6										
11	4.85	5.12	5.50	6.07	7.01	8.83	13.35	33.14	700.3									
12	4.71	4.91	5.19	5.58	6.17	7.15	9.03	13.75	34.62	763.9								
13	4.60	4.76	4.97	5.25	5.66	6.26	7.27	9.22	14.12	36.03	827.6							
14	4.51	4.64	4.81	5.02	5.32	5.73	6.35	7.39	9.40	14.48	37.40	891.3						
15	4.44	4.55	4.68	4.85	5.08	5.37	5.80	6.43	7.50	9.57	14.82	38.71	954.9					
16	4.38	4.48	4.59	4.72	4.90	5.12	5.43	5.86	6.51	7.60	9.73	15.15	39.98					
17	4.34	4.41	4.51	4.62	4.76	4.94	5.17	5.48	5.92	6.59	7.70	9.88	15.46	41.21				
18	4.30	4.36	4.44	4.54	4.66	4.80	4.98	5.21	5.53	5.98	6.66	7.80	10.03	15.76	42.41			
19	4.26	4.32	4.39	4.47	4.57	4.69	4.83	5.01	5.25	5.57	6.03	6.72	7.89	10.17	16.05			
20	4.23	4.29	4.35	4.42	4.50	4.60	4.72	4.86	5.05	5.29	5.62	6.08	6.79	7.98	10.31			
21	4.21	4.26	4.31	4.37	4.44	4.52	4.62	4.74	4.89	5.08	5.33	5.66	6.13	6.85	8.06			
22	4.19	4.23	4.28	4.33	4.39	4.46	4.55	4.65	4.77	4.92	5.11	5.36	5.70	6.18	6.91			
23	4.17	4.21	4.25	4.30	4.35	4.41	4.49	4.57	4.67	4.80	4.95	5.14	5.40	5.74	6.22	47.94		
24	4.15	4.19	4.22	4.27	4.32	4.37	4.43	4.51	4.59	4.70	4.82	4.98	5.17	5.43	5.78	17.36		
25	4.14	4.17	4.20	4.24	4.28	4.33	4.39	4.45	4.53	4.62	4.72	4.85	5.00	5.20	5.46	10.92		
26	4.12	4.15	4.18	4.22	4.26	4.30	4.35	4.41	4.47	4.55	4.64	4.74	4.87	5.03	5.23	8.43		

27	4.11	4.14	4.17	4.20	4.24	4.27	4.32	4.37	4.43	4.49	4.57	4.66	4.76	4.89	5.05	7.17		
28	4.10	4.13	4.15	4.18	4.21	4.25	4.29	4.33	4.38	4.44	4.51	4.59	4.68	4.78	4.91	6.43	52.90	
29	4.09	4.12	4.14	4.17	4.20	4.23	4.26	4.30	4.35	4.40	4.46	4.53	4.60	4.69	4.80	5.94	18.50	
30	4.09	4.11	4.13	4.15	4.18	4.21	4.24	4.28	4.32	4.36	4.42	4.47	4.54	4.62	4.71	5.60	11.44	
31	4.08	4.10	4.12	4.14	4.17	4.19	4.22	4.26	4.29	4.33	4.38	4.43	4.49	4.56	4.64	5.35	8.75	
32	4.07	4.09	4.11	4.13	4.15	4.18	4.21	4.24	4.27	4.31	4.35	4.39	4.45	4.50	4.57	5.16	7.40	
33	4.07	4.08	4.10	4.12	4.14	4.17	4.19	4.22	4.25	4.28	4.32	4.36	4.41	4.46	4.52	5.01	6.60	57.43
34	4.06	4.08	4.09	4.11	4.13	4.15	4.18	4.20	4.23	4.26	4.29	4.33	4.37	4.42	4.47	4.89	6.09	19.51
35	4.06	4.07	4.09	4.11	4.12	4.14	4.16	4.19	4.21	4.24	4.27	4.31	4.34	4.39	4.43	4.79	5.72	11.90
36	4.05	4.07	4.08	4.10	4.12	4.13	4.15	4.18	4.20	4.22	4.25	4.28	4.32	4.36	4.40	4.71	5.46	9.03
37	4.05	4.06	4.08	4.09	4.11	4.13	4.14	4.16	4.19	4.21	4.24	4.26	4.29	4.33	4.37	4.64	5.26	7.60
38	4.05	4.06	4.07	4.09	4.10	4.12	4.13	4.15	4.17	4.20	4.22	4.25	4.27	4.31	4.34	4.59	5.10	6.76
39	4.04	4.06	4.07	4.08	4.10	4.11	4.13	4.14	4.16	4.18	4.21	4.23	4.26	4.28	4.32	4.54	4.97	6.21
40	4.04	4.05	4.06	4.08	4.09	4.10	4.12	4.13	4.15	4.17	4.19	4.22	4.24	4.27	4.29	4.49	4.87	5.83
50	4.03	4.03	4.04	4.05	4.06	4.07	4.07	4.08	4.09	4.10	4.12	4.13	4.14	4.15	4.17	4.25	4.38	4.59
60	4.03	4.03	4.04	4.04	4.05	4.05	4.06	4.06	4.07	4.08	4.08	4.09	4.10	4.11	4.12	4.17	4.23	4.32
70	4.03	4.03	4.04	4.04	4.05	4.05	4.06	4.06	4.06	4.07	4.07	4.08	4.08	4.09	4.09	4.13	4.17	4.22
80	4.04	4.04	4.04	4.05	4.05	4.06	4.06	4.07	4.06	4.07	4.07	4.07	4.08	4.08	4.09	4.11	4.13	4.17
90	4.05	4.05	4.05	4.05	4.06	4.06	4.07	4.07	4.07	4.07	4.07	4.08	4.08	4.08	4.09	4.10	4.12	4.14
100	4.06	4.06	4.06	4.06	4.06	4.07	4.07	4.07	4.07	4.07	4.08	4.08	4.08	4.08	4.09	4.10	4.11	4.13
200	4.15	4.15	4.15	4.15	4.15	4.15	4.15	4.15	4.15	4.15	4.15	4.15	4.15	4.15	4.15	4.16	4.16	4.16
300	4.21	4.21	4.21	4.21	4.21	4.21	4.22	4.22	4.22	4.22	4.22	4.22	4.22	4.22	4.22	4.22	4.22	4.22
400	4.26	4.27	4.27	4.27	4.27	4.27	4.27	4.27	4.27	4.27	4.27	4.27	4.27	4.27	4.27	4.27	4.27	4.27
500	4.31	4.31	4.31	4.31	4.31	4.31	4.31	4.31	4.31	4.31	4.31	4.31	4.31	4.31	4.31	4.31	4.31	4.31

Table E Rankits for $n \leqslant 20$

Values of the rankits (or expected values of normal order statistics) not shown are found by symmetry. For example, the fifteenth largest rankit for a sample of size $n = 17$ is equal to the negative of the third rankit for $n = 17$, or 1.03.

					n					
i	1	2	3	4	5	6	7	8	9	10
1	0	−0.56	−0.85	−1.03	−1.16	−1.27	−1.35	−1.42	−1.49	−1.54
2		0.56	0.00	−0.30	−0.50	−0.64	−0.76	−0.85	−0.93	−1.00
3			0.85	0.30	0.00	−0.20	−0.35	−0.47	−0.57	−0.66
4				1.03	0.50	0.20	0.00	−0.15	−0.27	−0.38
5					1.16	0.64	0.35	0.15	0.00	−0.12
6						1.27	0.76	0.47	0.27	0.12

					n					
i	11	12	13	14	15	16	17	18	19	20
1	−1.59	−1.63	−1.67	−1.70	−1.74	−1.77	−1.79	−1.82	−1.84	−1.87
2	−1.06	−1.12	−1.16	−1.21	−1.25	−1.28	−1.32	−1.35	−1.38	−1.41
3	−0.73	−0.79	−0.85	−0.90	−0.95	−0.99	−1.03	−1.07	−1.10	−1.13
4	−0.46	−0.54	−0.60	−0.66	−0.71	−0.76	−0.81	−0.85	−0.89	−0.92
5	−0.22	−0.31	−0.39	−0.46	−0.52	−0.57	−0.62	−0.66	−0.71	−0.75
6	0.00	−0.10	−0.19	−0.27	−0.34	−0.40	−0.45	−0.50	−0.55	−0.59
7	0.22	0.10	0.00	−0.09	−0.17	−0.23	−0.30	−0.35	−0.40	−0.45
8	0.46	0.31	0.19	0.09	0.00	−0.08	−0.15	−0.21	−0.26	−0.31
9	0.73	0.54	0.39	0.27	0.17	0.08	0.00	−0.07	−0.13	−0.19
10	1.06	0.79	0.60	0.46	0.34	0.23	0.15	0.07	0.00	−0.06

Abridged with permission from Table 28 of E. S. Pearson and H. O. Hartley (1966), *Biometrika Tables for Statisticians*, Vol. 1, 3rd ed., London: Cambridge University.

REFERENCES

Acton, F. (1959). *Analysis of Straight Line Data*. New York: Dover.

Afifi, A. A. and R. M. Elashoff (1966). "Missing values in multivariate statistics I. Review of the literature." *J. Amer. Statist. Assoc.*, **61**, 595–604.

Aitken, A. C. (1934). "Note on selection from a multivariate normal population." *Proc. Edinburgh Math. Soc.*, **4**, 106–110.

Allen, D. M. (1974). "The relationship between variable selection and prediction." *Technometrics*, **16**, 125–127.

Allison, T. and D. V. Cicchetti (1976). "Sleep in mammals: Ecological and constitutional correlates." *Science*, **194**, 732–734.

Andrews, D. F. (1971). "Significance testing based on residuals." *Biometrika*, **58**, 139–148.

Andrews, D. F. (1974). "A robust method for multiple linear regression." *Technometrics*, **16**, 523–531.

Andrews, D. F., P. Bickel, F. Hampel, P. Huber, W. Rogers, and J. W. Tukey (1972). *Robust Estimates of Location*. Princeton, N.J.: Princeton University.

Anscombe, F. J. (1973). "Graphs in statistical analysis." *Amer. Statist.*, **27**, 17–21.

Atkinson, A. C. (1973). "Testing transformations to normality." *J. Roy. Statist. Assoc., Ser. B*, **35**, 473–479.

Baes, C. F. and H. H. Kellogg (1953). "Effect of dissolved sulphur on the surface tension of liquid copper." *J. Metals*, **5**, 643–648.

Barnett, V. and T. Lewis (1978). *Outliers in Statistical Data*. Chichester: Wiley.

Beale, E. M. L., M. G. Kendall, and D. W. Mann (1967). "The discarding of variables in multivariate analysis." *Biometrika*, **54**, 537–566.

Beale, E. M. L. and R. J. Little (1975). "Missing values in multivariate analysis." *J. R. Statist. Soc. Ser. B*, **37**, 129–145.

Berger, J. (1975). "Minimax estimation of location vectors for a wide variety of densities." *Ann. Statist.*, **3**, 1318–1328.

Berk, K. N. (1977). "Tolerance and condition in regression computations." *J. Amer. Statist. Assoc.*, **72**, 863–866.

Bingham, C. and K. Larntz (1977). "Comment on 'A simulation study of alternatives to ordinary least squares.'" *J. Amer. Statist. Assoc.*, **72**, 97–102.

Bland, Johnny (1978). "A comparison of certain aspects of ontogeny in the long and short shoots of McIntosh apple during one annual growth cycle." Unpublished Ph.D. Thesis, University of Minnesota, St. Paul, Minnesota.

Blom, G. (1958). *Statistical Estimates and Transformed Beta Variates.* New York: Wiley.

Box, G. E. P. and D. R. Cox (1964). "An analysis of transformations (with discussion)." *J. R. Statist. Soc. Ser. B*, **26**, 211–246.

Box, G. E. P., J. S. Hunter, and W.G. Hunter (1978). *Statistics for Experimentors.* New York: Wiley.

Box, G. E. P. and P. W. Tidwell (1962). "Transformation of the independent variables." *Technometrics*, **4**, 531–550.

Box, G. E. P. and K. B. Wilson (1951). "On the experimental attainment of optimal conditions." *J. R. Statist. Soc. Ser. B*, **13**, 1–45.

Buck, S. F. (1960). "A method of estimating missing values in multivariate data suitable for use with an electronic computer." *J. R. Statist. Soc. Ser. B*, **22**, 302–306.

Burt, C. (1966). "The genetic determination of differences in intelligence: A study of monozygotic twins reared together and apart." *Brit. J. Psych.*, **57**, 137–153.

Chambers, John M. (1977). *Computational Methods for Data Analysis.* New York: Wiley.

Chapman, H. M. and D. B. Demeritt (1936). *Elements of Forest Mensuration*, 2nd ed., Albany, N.Y.: Williams Press.

Cook, R. D. (1977). "Detection of influential observations in linear regression." *Technometrics*, **19**, 15–18.

Cook, R. D. (1979). "Influential observations in linear regression." *J. Amer. Statist. Assoc.*, **74**, 169–174.

Cook, R. D. and J. O. Jacobson (1978). "Analysis of 1977 West Hudson Bay snow goose surveys." Unpublished report, Canadian Wildlife Service.

Cook, R. D. and S. Weisberg (1979). "Use of the minimum covering ellipsoid in regression," University of Minnesota Technical Report.

Cook, R. D. and S. Weisberg (1980). "Characterizations of an empirical influence function for detecting influential cases in regression." *Technometrics*, **22**, in press.

Cox, D. R. (1958). *The Planning of Experiments.* New York: Wiley.

Cox, D. R. (1970). *The Analysis of Binary Data.* London: Methuen.

Daniel, C. (1976). *Applications of Statistics to Industrial Experiments.* New York: Wiley.

Daniel C. and F. Wood (1971). *Fitting Equations to Data.* New York: Wiley.

Dempster, A. P. (1973). "Alternatives to least squares in multiple regression." In D. G. Kale and R. P. Gupta (Eds.), *Multivariate Statistical Inference.* Amsterdam: North Holland.

Dempster, A. P., N. M. Laird, and D. B. Rubin (1977). "Maximum likelihood from incomplete data via the EM algorithm (with discussion)." *J. R. Statist. Soc. Ser. B*, **39**, 1–38.

Dempster, A. P., M. Schatzoff, and N. Wermuth (1977). "A simulation study of alternatives to ordinary least squares (with discussion)." *J. Amer. Statist. Assoc.*, **72**, 77–104.

Dixon, W. J. and M. Brown (1979). *BMDP Biomedical Computer Programs, 1979*. Berkeley: University of California.

Dongarra, J., J. P. Bunch, C. B. Moler, and G. W. Stewart (1979). *The LINPACK Users Guide*. Philadelphia: SIAM.

Draper, N. and W. G. Hunter (1969). "Transformations: Some examples revisited." *Technometrics*, **11**, 23–40.

Draper, N. and H. Smith (1966). *Applied Regression Analysis*. New York: Wiley.

Draper, N. R. and R. C. Van Nostrand (1978). "Ridge regression: Is it worthwhile." Technical Report No. 501, Dept. of Statistics, University of Wisconsin.

Draper, N. R. and R. C. Van Nostrand (1979). "Ridge regression and James-Stein estimation: Review and comments." *Technometrics*, **21**, 451–466.

Durbin, J. and G. S. Watson (1950). "Testing for serial correlation in least squares regression I." *Biometrika*, **37**, 409–428.

Durbin, J. and G. S. Watson (1951). "Testing for serial correlation in least squares regression II." *Biometrika*, **38**, 159–178.

Durbin, J. and G. S. Watson (1971). "Testing for serial correlation in least squares regression III." *Biometrika*, **58**, 1–19.

Efron, B. and C. Morris (1973). "Stein's estimation rule and its competitors—an empirical Bayes approach." *J. Amer. Statist. Assoc.*, **68**, 117–130.

Efron, B. and C. Morris (1975). "Data analysis using Stein's estimator and its generalizations." *J. Amer. Statist. Assoc.*, **70**, 311–319.

Ezekiel, M. and F. A. Fox (1959). *Methods of Correlation and Regression Analysis*. New York: Wiley.

Federov, V. V. (1972). *Theory of Optimal Experiments*. New York: Academic.

Fienberg, S. E. (1977). *The Analysis of Cross Classified Categorical Data*. Cambridge, Mass.: MIT.

Forbes, James D. (1857). "Further experiments and remarks on the measurement of heights by the boiling point of water." *Trans. R. Soc. Edinburgh*, **21**, 135–143.

Freeman, M. F. and J. W. Tukey (1950). "Transformations related to the angular and the square root." *Ann. Math. Statist.*, **21**, 607–11.

Furnival, G. and R. Wilson (1974). "Regression by leaps and bounds." *Technometrics*, **16**, 499–511.

Garside, M. J. (1971). "Some computational procedures for the best subset problem." *Appl. Statist.*, **20**, 8–15.

Gnanadesikan, R. (1977). *Methods for Statistical Analysis of Multivariate Data*. New York: Wiley.

Goldstein, M. and A. F. M. Smith (1974). "Ridge type estimators for regression analysis." *J. R. Statist. Soc. Ser. B*, **36**, 284–291.

Graybill, F. A. (1969). *Introduction to Matrices with Statistical Applications*. Belmont, Calif.: Wadsworth.

Hald, A. (1960). *Statistical Theory with Engineering Application*. New York: Wiley.

Hampel, F. (1974). "The influence curve and its role in robust estimation." *J. Amer. Statist. Assoc.*, **69**, 383–393.

Hartley, H. O. and R. R. Hocking (1971). "The analysis of incomplete data." *Biometrics*, **27**, 783–808.

Hinkley, D. V. (1977). "Jackknifing in unbalanced situations." *Technometrics*, **19**, 285–92.

Hoaglin, D. C. and R. Welch (1978). "The hat matrix in regression and ANOVA." *Amer. Statist.*, **32**, 17–22.

Hocking, R. R. (1976). "The analysis and selection of variables in linear regression." *Biometrics*, **32**, 1–40.

Hocking, R. R. and R. N. Leslie (1967). "Selection of the best subset in regression analysis." *Technometrics*, **2**, 531–540.

Hodges, S. D. and P. G. Moore (1972). "Data uncertainties and least squares regression." *Appl. Statist.*, **21**, 185–195.

Hoerl, A. E. and R. W. Kennard (1970a). "Ridge regression: biased estimation for nonorthogonal problems." *Technometrics*, **12**, 55–67.

Hoerl, A. E. and R. W. Kennard (1970b)."Ridge regression: Applications to nonorthogonal problems." *Technometrics*, **12**, 69–82.

Huber, P. J. (1973). "Robust regression: Asymptotics, conjectures and Monte Carlo." *Ann. Statist.*, **1**, 799–821.

Huber, P. J. (1977). *Robust Statistical Procedures*, No. 27, Regional Conference Series in Applied Mathematics. Philadelphia: Society for Industrial and Applied Mathematics.

IMSL (1979). *The IMSL Library*. Houston: International Mathematical Statistical Library.

James, W. and C. Stein (1961). "Estimation with quadratic loss," in *Proceedings of the Fourth Berkeley Symposium on Mathematical Statistics and Probability,* Vol. 1. Berkeley: University of California, 361–379.

Jensen, Ronald (1977). "Evinrude's computerized quality control productivity." *Qual. Prog.*, **10**, 12–16.

John, P. W. M. (1971). *Statistical Design and Analysis of Experiments*. New York: Macmillan.

Johnston, J. (1963). *Econometric Methods*. New York: McGraw-Hill.

Kennedy, W. J. and T. A. Bancroft (1971). "Model building for prediction in regression based on repeated significance tests." *Ann. Math. Statist.*, **42**, 1273–1284.

LaMotte, L. R. and R. R. Hocking (1970). "Computational efficiency in the selection of regression variables." *Technometrics*, **12**, 83–93.

Land, C. E. (1974). "Confidence interval estimation for means after data transformations to normality." *J. Amer. Statist. Assoc.* **69**, 795–802 (Correction, *ibid*, **71**, 255).

Larsen, W. A. and S. A. McCleary (1972). "The use of partial residual plots in regression analysis." *Technometrics*, **14**, 781–790.

Lawley, D. N. (1943). "A note on Karl Person's selection formulae." *Proc. R. Soc. Edinburgh*, **62**, 28–30.

Lawson, C. L. and R. J. Hanson (1974). *Linear Least-Squares Problems*. Englewood Cliffs, N.J.: Prentice-Hall.

Lindgren, B. W. (1976). *Statistical Theory*, 3rd ed., New York: McMillan.

Little, R. J. A. (1979). "Maximum likelihood inference for multiple regression with missing values: A simulation study." *J. R. Statist. Soc. Ser. B*, **41**, 76–87.

Longley, J. W. (1967). "An appraisal of least squares programs for the electronic computer from the point of view of the user." *J. Amer. Statist. Assoc.*, **62**, 819–841.

Lund, Richard E. (1975). "Tables for an approximate test for outliers in linear regression." *Technometrics*, **17**, 473–476.

Mallows, C. L. (1973). "Some comments on C_p." *Technometrics*, **15**, 661–676.

Mansfield, E. R., J. T. Webster, and R. F. Gunst (1977). "An analytic variable selection technique for principal component regression." *Appl. Statist.*, **26**, 34–40.

Mantel, N. (1970). "Why stepdown procedures in variable selection." *Technometrics*, **12**, 621–25.

Marler, G. D. (1969). *The Story of Old Faithful*. W. Yellowstone, Wyoming: Yellowstone Library and Museum Association.

Marquart, D. W. (1970) "Generalized inverses, ridge regression and biased linear estimation." *Technometrics*, **12**, 591–612.

Marquart, D. W. and R. E. Snee (1975). "Ridge regression in practice." *Amer. Statist.*, **12**, 3–19.

Miller, Rupert (1965). *Simultaneous Inference*. New York: McGraw-Hill.

Miller, Rupert (1977). "Developments in multiple comparisons 1966–76." *J. Amer. Statist. Assoc.* **72**, 779–88.

Moore, J. A. (1975). "Total Biochemical Oxygen Demand of Animal Manures." Unpublished Ph.D. dissertation, University of Minnesota.

Morgan, J. A. and J. F. Tartar (1972). "Calculation of the residual sum of squares for all possible regressions." *Technometrics*, **14**, 317–25.

Morrison, D. F. (1977). *Multivariate Statistical Methods*, 2nd ed. New York: McGraw-Hill.

Mosteller, F. and J. W. Tukey (1977). *Data Analysis and Linear Regression*. Reading, Mass.: Addison-Wesley.

Myers, R. H. (1971). *Response Surface Methodology*. Boston: Allyn and Bacon.

Narula, S. C. and J. F. Wellington (1977). "Prediction, linear regression and minimum sum of relative errors." *Technometrics*, **19**, 185–90.

274 References

Obenchain, R. L. (1975). "Ridge analysis following a preliminary test of the shrunken hypothesis." *Technometrics*, **17**, 431–45.

Orchard, T. and M. A. Woodbury (1972). "A missing information principle: Theory and applications," in *Proceedings of the Sixth Berkeley Symposium on Mathematical Statistics and Probability*, Vol. 1. Berkeley: University of California, 697–715.

Pearson, K. (1914). *Life and Letters and Labours of Francis Galton*. Cambridge: University Press.

Plackett, R. L. (1950). "Some theorems in least squares." *Biometrika*, **37**, 149–57.

Press, S. J. (1972). *Applied Multivariate Analysis*. New York: Holt, Rinehart and Winston.

Renshaw, E. (1958). "Scientific appraisal." *Nat. Tax J.*, **11**, 314–22.

Rolph, J. E. (1976). "Choosing shrinkage estimators for regression problems." *Commun. Statist.—Theor. Meth.*, **A5**, 789–802.

Rubin, D. B. (1974). "Characterizing the estimation of parameters in incomplete data problems." *J. Amer. Statist. Assoc.*, **69**, 467–74.

Rubin, D. B. (1976). "Inference and missing data." *Biometrika*, **63**, 581–92.

Rubin, D. B. (1980). "Using empirical Bayes techniques in the Law School validity studies." *J. Amer. Statist. Assoc.* (in press).

Saw, J. G. (1966). "A conservative test for the concurrence of several regression lines and related problems." *Biometrika*, **53**, 272–75.

Schatzoff, M., S. Fienberg, and R. Tsao (1968). "Efficient calculation of all possible regressions." *Technometrics*, **10**, 769–79.

Scheffe, H. (1957). *The Analysis of Variance*. New York: Wiley.

Sclove, S. (1968). "Improved estimators for coefficients in linear regression." *J. Amer. Statist. Assoc.*, **63**, 596–606.

Sclove, S. (1972). "(Y vs. X) or (log Y vs. X)?" *Technometrics*, **14**, 391–403.

Searle, S. R. (1966). *Matrix Algebra for the Biological Sciences*. New York: Wiley.

Seber, G. A. F. (1977). *Linear Regression Analysis*. New York: Wiley.

Silvey, S. D. (1969). "Multicollinearity and imprecise estimation." *J. R. Statist. Soc. Ser. B.*, **31**, 539–52.

Smith, G. and F. Campbell (1980). "A critique of some ridge regression methods (with discussion)." *J. Amer. Statist. Assoc.*, **75**, 74–103.

Stein, C. (1956). "Inadmissibility of the usual estimator for the mean of a multivariate normal distribution," in *Third Berkeley Symposium on Probability and Statistics*, Vol. 1. Berkeley: University of California.

Stewart, G. W. (1974). *Introduction to Matrix Computations*. New York: Academic.

Stewart, G. W. (1979). "Assessing the effects of variable error in linear regression." Computer Science Technical Report No. 818, University of Maryland.

Strawderman, W. (1978). "Minimax adaptive generalized ridge regression estimators." *J. Amer. Statist. Assoc.*, **73**, 623–27.

Thisted, R. (1978a). "On generalized ridge regression," Technical Report Number 57, Dept. of Statistics, University of Chicago.

Thisted, R. (1978b). "Multicollinearity, information and ridge regression," Technical Report Number 58, Dept. of Statistics, University of Chicago.

Titterington, D. M. (1975). "Optimal design: some geometric aspects of D-optimality." *Biometrika*, **62**, 313–320.

Titterington, D. M. (1978). "Estimation of correlation coefficients by elliptical trimming." *Appl. Statist.*, **27**, 227–234.

Tukey, J. W. (1949). "One degree of freedom for non-additivity." *Biometrics*, **5**, 232–42.

Tuddenham, R. D. and M. M. Snyder (1954). "Physical growth of California boys and girls from birth to age 18." *Calif. Publ. Child Develop.*, **1**, 183–364.

Weisberg, H., E. Beier, H. Brody, R. Patton, K. Raychaudhari, H. Takeda, R. Thern, and R. Van Berg (1978). "*s*-dependence of proton fragmentation by hadrons. II. Incident laboratory momenta 30–250 GeV/c." *Phys. Rev. D*, **17**, 2875–2887.

Weisberg, S. (1979). "A statistic for allocating C_p to individual cases," unpublished.

Weisberg, S. and C. Bingham (1975). "An analysis of variance test for normality suitable for machine calculation." *Technometrics*, **17**, 133.

Wilk, M. B. and Gnanadesikan, R. (1968). "Probability plotting methods for the analysis of data." *Biometrika*, **55**, 1–17.

Wilm, H. G. (1950). "Statistical control in hydrologic forecasting." *Res. Notes*, **61**, Pacific Northwest Forest Range Experiment Station, Oregon.

Wood, F. S. (1973). "The use of individual effects and residuals in fitting equations to data." *Technometrics*, **15**, 677–95.

Woodley, W. L., J. Simpson, R. Biondini, and J. Berkeley (1977). "Rainfall results 1970–75: Florida area cumulus experiment." *Science*, **195**, 735–42.

SYMBOL INDEX

The page references give the location of the first occurrence of a symbol or the place where it is defined. If more than one page reference is given, then the symbol may have more than one meaning, in which case the proper definition should be clear from context, or else the symbol is defined with varying levels of generality in various places. Symbols beginning with Greek letters are listed after the Roman letter symbols.

INDEX